A History of

Our Emerging

Consciousness

A process of becoming

A History of

Our Emerging Consciousness

A process of becoming

by

Andrew Bradbery

Published by Historic Milestones Ltd

To Elisabeth and David.

First Published in June 2015.
Published by Historic Milestones Ltd.
May 2016.
ouremergingconsciousness.com

ISBN 978-0-9934559-2-6

"Behold the image of the end
in the mirror of the beginning"

Rumi: Mathnawi.IV.

Contents

Part 2

Building and Testing a
Model for Our Emerging Consciousness

Acknowledgements

The completion of this book has left me with a deep appreciation of the many ways I have been helped to reach this conclusion. During the early years of this work, the following people were a great source of help and encouragement, Bars Bullen, Sally Sadler, Jennette Roof, Jane Walters, and Sue Whitton. In the years that followed the skills of Jane Symons, Sally Tolhurst, Nicki Hayes, and Jenny Chantler assisted me with numerous chapters. In particular, I want to thank Helen Taylor and Clare Higson for their years of help during much of the projects duration. The help from Liz Bradbery, Rowena Gavars, Deborah Stratford, Steve Bonney and Mark Taylor is much appreciated. Credits to Andrew Akehurst's Community Workshop for the cover design, printing services, Fig 8.1, Fig 17 and Fig 18. Credits to Tig Heffernan for Fig 6.1 and Fig 8.2.

These thanks must be extended to include the many authors whose published work provided the wealth of reference material used in this book. This vast resource was vital to research and present the overview, which this work introduces. For evidence about many past cultures including the thirty millennia of Stone Age periods, my particular appreciation is extended to the scholars referred to in the text.

The efficient service I received from my local library for researching many of the subjects considered is appreciated.

Preface

It seems to me that much of the modern world's suffering arises from failures to use our collective intelligence wisely. The idea of men playing god suggests an ongoing struggle that has been increasingly destroying our planet. The environmental changes that people are causing seem out of our control.

The first aim of this book is to identify new clues able to help us better understand what is happening. The second is to see how the growth of intelligence somehow relates to other qualities of consciousness that have also emerged during past millennia.

Introduction

Optimists viewing our modern culture are likely to be continually amazed by the capabilities being expressed in countless spheres of endeavour. There seems no limit can be placed on human creativity. Such ideas are believed by many realists to be fantasies in view of the global scale of the challenges facing us. The consequences of the consumer's often-unbridled consumption are increasingly threatening our prospects and those of future generations. The demands growing populations' place on natural resources and the expansion of industry are destroying our planet's capacity to support us. What can this duality teach us about the inadequacies of our collective intelligence?

To understand the continuing growth of these skills better we will consider their progress over past millennia. This work considers historical and archaeological evidence to create an overview of the growth of intelligence. We look at the sequence and ways its physical, emotional, mental and spiritual dimensions have developed since the last Ice Age. When we look at these the emergence of other qualities of consciousness is identified. Each of these has their own physical, emotional, mental and spiritual phases of expression. These qualities of consciousness enhanced the expressions of intelligence that emerged as numerous past cultures refined their abilities to meet their survival challenges.

The timescale of these developments suggest their correspondence with the qualities described by a little understood natural cycle. This forms a basis for a fascinating multi-layered model of our emerging consciousness for us to assess. To test its validity beyond the study of intelligence the model is extended. It considers our modern understandings of the prehistoric evidence arising from the early revolution that introduced the human imagination.

From this period, we trace how other qualities of consciousness also emerged in physical, emotional, mental and spiritual phases. This hypothesis describes a continuum, which includes the opportunities this natural cycle is now providing for us.

Part One

Qualities of

Our Emerging Intelligence

1

Shifting Sands

Many modern cultures accept that the trained intellect is a resource that potentially increases the ability of people to make valued contributions to their society. It is appreciated that the trained intellect can provide the abilities needed to achieve personal ambitions. This quest is vital for many of the highest cultural accolades.

However, it does risk parents sacrificing much of their children's fun and freedom for the promise of dazzling prizes during later life. Although this may make sense in upholding the fabric of modern life, when it comes to our really important life choices and challenges, the intellect is usually irrelevant and redundant. It is at a loss when confronted with the chaos and stress life often presents us with, when answers elude us.

Much of what is most significant in human life does not 'make sense'. Do we choose our homes, food, loves, friends, partners, clothes, or our jobs solely because they 'make sense'? Perhaps in part – we are at times capable of making logical, rational choices. However, pause for a moment and remember the landmarks of our lives. How much did they have to do with the intellect's logic and rationality? None! The intellect alone (often confused with intelligence, which will be discussed later) cannot give us a good life – the kind of happiness we seek – either individually or collectively.

The human condition has been examined minutely and described throughout time. In classical literature, we find at one extreme the flaws and frailties described as the seven deadly sins. In contrast at the opposite polarity, sacred writing such as the Christian Beatitudes, express how different strands of spiritual motivation bring blessings. When humans experience any of these extremes, what role does intellectual excellence play? Is the intellect anything other than the servant of whatever personal motivation, exalted or base is driving the individual? Human aspiration has by definition, a target. Choices are made. However, since life is always recreating itself, the mistakes from the past are constantly reappearing in new disguises returning repeatedly until changes in consciousness correct the errors.

Many historic choices made to satisfy human need (and greed) were only realised by overpowering the natural world. This strategy, driving the greater global exploitation of natural resources, is increasingly recognised as leading to ecological disaster. The human activities still taking place and contributing to the looming catastrophes are sobering guides as to what actions to avoid if we are to secure a future of benefit to all. A new relationship of international stewardship based on cooperation is struggling to become accepted as a necessity to replace the excessive consumption fostered by greed. History provides a huge resource from which we might learn from past mistakes. There are myriad examples of civilizations prospering from changes in their traditional cultural practices. For example, the advent of farming provided alternative sources of food necessary for populations to increase. The climate change that ended the last Ice Age, releasing the fertility of the formerly frozen environments and dramatically increasing the capability of many regions to grow food, introduces a field of study with its own complex history. A greater understanding of the timing and the limits within which various aspects of human life have changed will yield insights into some of the most far-reaching changes in human history that have brought us to where we are today.

However, a key question is this: is it natural processes or

random human activities that drive the changes in human consciousness that lead to such watershed changes? This book searches for a model that describes *the process underlying the continual emergence of new human capabilities.*

From what we know of Stone Age life, it is clear how different we are today. Being human now depends on learning from the experiences of our ancestors. These shaped the cultural developments that took place during scores of prehistoric millennia. The changes in human consciousness that have taken place during these ages are the subject of this study, and one objective has been to bring a new sense of order into understanding our past.

The different periods traditionally used to identify times of significant change; the Stone, Bronze and Iron Ages saw the development of the tools that eventually aided the birth of the Industrial Revolution and modern technology. The secular initiatives taken during the centuries after the Renaissance introduced the phenomenon of industrial power together with its associated social repercussions. As these changes developed, they increasingly threatened the two bastions of traditional power, the Church and the landowners. The extremes of industrial wealth created during the Victorian Age clearly illustrated the social cost paid. The dark satanic mills that forged many industrial workers could not be ignored. A few generations earlier, the Church and other landowners hid deprivation of the poor away in villages where the traditional seats of power were shared. These had long maintained the status quo with the fear of God, homelessness and not being able to feed one's children. The long-established power of those revered stewards of all that was held sacred in society and as of greatest value, those arbiters of what was deemed holy, and owners of precious acres began to wane.

The new financial power wielded by industrialists was not locked in scripture or land. Its mutability allowed wealth to be more widely dispersed. Power began to trickle down the social scale and gradually lessen the fears of the powerless. Through hard won incremental changes, the new freedom allowed a sense of the sacred to emerge in fresh ways. A duality between the religious and the personal came into sharper focus.

In the westernised mind, the term 'sacred' is generally connected with the people, events and paraphernalia of organised religion: most usually, perhaps, a conventional conservative expression of a faith. The broader dictionary definition includes aspects such as 'beyond criticism', 'not to be violated', 'dedicated to the exclusive use of', and so on, yet it is the religious ownership of the word 'sacred' that holds the strongest association in the westernised mind.

Our understanding of the sacred may include the highly valued experience of reverence arising for someone or something. At the other end of the spectrum, it may focus on the mystical: gnostic experience of a divine source of love, joy, peace or power. From the cradle to the grave experiences arise that may bring feelings of awe that could be deemed sacred. The point to be made here is that the process whereby sacred personal experiences are created is subjective and individual.

As we have said the term 'sacred' primarily refers only to the people, events and paraphernalia that enrich the world's religions. Holy relics, sites, buildings, books, images and music have become enshrined as part of the identity of each new generation of believers. For many people these may be a combination of vital physical, emotional, mental and spiritual expressions of their culture and are protected to maintain the stability of the unique beliefs that constitute its foundations.

Religious and political education provides information about different cultures and faiths, and the results of the decision-making process of those in authority who decide what others must accept as sacred. This may have nothing to do with the individual's personal witness to the experience of the sacred arising from within. The life-enriching potential of a mystical experience of divine love, joy or peace can be deflected by the theological details that divide the religions of the world: it could be said that the devil is in such details.

Before the world's major monotheistic religions became established many cultures had their pagan spiritual beliefs centred on animal and bird fetishes or numerous gods and goddesses. In this work, it is presumed that the worship of any 'deity' by any person at any time may have been motivated by the aspiration to experience non-separation with the sacred in their life. In addition, many non-religious aspects of life are held sacred by countless millions. These may be such things as the physical beauty of natural shapes, the emotions that bind people together, the mental understanding provided by knowledge, or the spiritual inspiration that changes a painful personal experience into one that is deeply valued. The religious believer will be able to interpret these as God's omnipotence enabling the experience of the sacred to be evoked in the mundane, whereas the non-religious will view them as entirely unconnected with any religious context.

Irrespective of the manner in which the sacred is sought, who is not on the quest to experience consciously or unconsciously the sacred in their unique way? When realised, is this not just a mirror of their personal sacredness?

As the holders of religious power no longer have the monopoly on judgments as to the 'sacredness' of any subject there is no universal formula for the kind of experience able to evoke within the individual an awareness of the sacred. Any physical, emotional, mental or spiritual sphere of life bringing a peak of cultural excellence has the potential to be deemed sacred. Never before has there existed such a wide spectrum of opportunity for sacred experiences, which may also lead to one's own highest form of creative expression.

The emotions that drive the individual's search for the

sacred often have a magnetic power that influences and attracts other seekers. In this way, sub-cultures may be propagated. Their potential to flower or decay depends on many factors, each with their own sacred agendas. Two of these near the top of the hierarchy of the sacred concern the security needed for a population to thrive and the measures taken to promote ethical standards of personal and institutional behaviour. At different times, different nations adopted different methods to nurture different expressions of the sacred.

Many phenomena once accepted as sacred within a culture are subsequently replaced by later generations with the power to adopt new gods, goddesses or material idols for veneration and worship. A few of these, which will be considered later, arose from the inspired creativity of 'sages' recognised by elite groups within a society as contributors to sacred knowledge. A look into history reveals whether or not the sages' contribution to the lives of those around them resulted in its widespread adoption to serve the common good. In Chapters 3 to 6, we explore some of the beneficial ideas that proved so conducive to serving the common good that they are now intrinsic to culture on a global scale and exemplify aspects of physical, emotional, mental and spiritual intelligence in familiar aspects of life. Again, religious believers are able to interpret each expression of progress as a gift from God, whereas the non-religious will necessarily interpret them differently.

The nature of the contribution, perhaps originally regarded as sacred by the originator, as a means to improve people's lives, may be lost or overridden if hi-jacked by a self-serving and powerful minority. The historical record shows us a multitude of religious, political, industrial and economic initiatives in which vested interests were favoured. Despite this imperative to maintain the status quo many historians believe that over the millennia different cultures, through their quests for excellence, have contributed specific steps that seem (until recently) to be taking successive civilizations forward in some progressive way.

As science has revealed more of nature's secrets the principle of correspondence is increasingly appreciated as echoing throughout nature's wholeness. This arises from an understanding that the Universe emanates from one source with common laws and characteristics that apply to each aspect of creation manifesting in its own plane of expression.

The various strands of creative originality, which successive civilisations have spun and woven into the rich tapestry of human possibilities, have not yet completed an image. Moreover, there is no consensus as to the existence of such a final image - let alone as to its nature.

Over the ages each new strand of the tapestry manifested by a sage within a particular culture has continually redefined what it can mean to be human and is perhaps, more sacred than that which has generally been deemed so. The refinements made within different cultures, to broaden the spectrum of the individual's choice of sacred experiences reveal a history of our emerging consciousness that this book will explore.

.

2

Our Changing Minds

During their first years, infants learn what it means to be alive in a physical body. Blessed with good health their priority is to extend the possibilities of what it means to be alive. What they achieve depends on many factors not least the family traits passed unconsciously from generation to generation. The experiences of their limitations challenge them to learn to survive within their own little world. The opportunities provided by the different environments experienced test the expression of their spirit. Can we overestimate the influence the strengths and weaknesses their early teachers have on this motivation to become free of un-necessary limitations? Later as a young boy or girl, the limiting beliefs and fears imposed by cultural conditioning do not have to define the extent of their ambitions. The modern environment presents young minds with an unprecedented range of stimuli.

To extend the range of physical accomplishments, sports continually test the body's limits of what is humanly possible. The challenge to follow the necessary disciplines of training and diet may later impose unacceptable demands on their social life and finances. For those determined to be world record breakers their motivation has to be sufficient to accept any number of privations as the focus on their objective becomes more intense. These physical ambitions have other related personal emotional, mental and spiritual challenges.

The resistance to gender equality inherent within patriarchal cultures illustrates emotional factors affecting spirit. This is one example of the wider challenge to acknowledge the care needed to address thoroughly the deep cultural differences that exist across our world. When we consider the idea of spiritual limitations imposed on the expression of the human spirit we find that many religious traditions still nurture fear to discourage believers from moving beyond accepted cultural norms. Ignorance of personal fulfilment or the exploration of the spiritual life is magnified by the belief (encouraged by the advertising industry) that lasting fulfilment will be found in materialistic pursuits.

Evidence of the consequences of cultures intent on propagating material consumption without controlling their debt; illustrate the choices that led to the 2008 credit crisis and the resulting global economic depression. Collectively they have contributed to undermining our will to support the people of the under developed nations, whilst increasingly destroying our planet's capacity to support human life. These changes are aggravated by the many organisations with huge skill resources devoted to increasing their power. It seems little can be done to improve ecological concerns. In first world markets, this influence promotes social separation as expressed by the status of exclusivity encouraged by marketing. The objective of expanding or maintaining control, based on consumption, competition and economic growth is not yet generally threatened by consumers with ethics at the heart of their ecological interests. While an understanding of the power of vested interests is growing within the ranks of the educated, consumers in the third world are powerless against this ruthlessness. This imbalance underlines how much the modern world needs the intelligence to include spiritual dimensions.

The challenges faced by individuals when deciding which cognitive skills will serve them best in these testing times are not easily resolved. The cultural diversity of the global village illustrates the powerful forces the modern mind has to wrestle with. The challenge to face and assess practical choices is compounded by the increasing complexity arising from the growth of intelligence during the last few centuries. During this time, the intellect has been successfully refined to the point where humankind has such power that our species has near supremacy over all life on Earth. Fundamental to this development has been the capacity of the mind to make new physical, emotional, mental and spiritual connections, to discern and remember the useful ones, and to communicate this experience to others. Countless inventions vital to the modern world followed the growth of education. Great inventors, inspired by new ideas, aspired to introduce all manner of machines capable of performing useful functions. The Industrial Revolution was driven by the power of the steam engine. Where would we be without computers today? Good ideas catch on, often saving time and money.

Over the centuries, the process of making different types of connections has given billions of people more power when they learn the needed skills to empower themselves. However, the results of all this 'intellectual progress' often testify to how little intelligence was and still is being expressed.

The power of the intellect that may arise from the successful honing of mental skills has established a priority that is central to modern life and given to each new generation: namely to exercise ever-greater intellectual skill. This seems a legitimate cultural goal. However, increasing numbers of parents and teachers believe that this priority to maximise the intellectual skills of schoolchildren is not an adequate foundation for their future happiness. This shortcoming is just one of the challenges increasingly recognised, namely that the intellect may or may not express much intelligence. While the intellect may be a powerful tool for us to use, the motivation, which drives our ambition, introduces the matter of ethical behaviour: *intelligence has a spiritual dimension.*

An awareness of three spiritual cultures emerged around 2,500 years ago when the philosophies of Pythagoras, Buddha and Confucius shaped three different ways of life. Each identified and focused on a sacred aspect of life's mysteries: to understand them, to be part of them or to live in harmony with them. They each illustrate a different approach to how the individual may consciously change their mind to enhance their life by using different sets of cognitive skills recognised by their culture as contributing to and supporting its priorities. The challenge to create a culture embodying the complementary aspects of these three philosophies is yet to be recognised.

Many millennia before the human mind was introduced to the wisdom provided by the above three sages the greater potential of the mind emerged. This created the c. 40,000 BC start of the Upper Palaeolithic Revolution and changed Stone Age life forever. To widen our perspective on this matter we look for insights into these prehistoric changes of the mind archaeology provides for us. This rich source of information includes research about how our ancestors began the simultaneous use of several abilities. The archaeologist Prof. Steven Mithen from Reading University uses the term cognitive fluidity in his title, *The Prehistory of the Mind*. In his theory for the evolution of the mind, he reviews the work of many respected psychologists and considers their contributions in his study of the mind. Prof. Mithen argues that the Stone Age mind had technical, social and natural history intelligence existing independently within its own domains. He uses the symbolism of the separate tools on a Swiss Army knife and their selection for specialised tasks.

Prof. Mithen considers in Chapter 9 of the above book the origins of art and religion and identifies the cognitive fluidity between the above three types of intelligence needed for artistic expression. At the start of the Upper Palaeolithic technical, social and natural history intelligence began to work together and express artistic abilities.[1]

The practical implications of this new Stone Age expression of creativity are extensive and far-reaching. Describing these,

---

[1] Steven Mithen, *The Prehistory of the Mind*, p. 162.

Prof. Sir Paul Mellars from Cambridge University writes about the Aurignacian technology, c. 35,000 – 29,000 BC: the first major stage at the start of the Upper Palaeolithic. Although the production of stone tools changed at this time, the most dramatic development was the carving of designed bone, antler and ivory artefacts'.[2] The term 'technology' describes a systematic use of knowledge to manufacture artefacts copied from artistic originals used within a culture with the spiritual, mental, emotional and physical abilities to express it.

These skills provided the basis for *Homo sapiens* to continue the long process of becoming more proficient when presented with new ordeals. This challenge reached a peak at the last Ice Age's last glacial maximum (LGM) about 20,000 years ago. As the cold intensified, c. 27,000 BC, statuettes and images of the Earth Goddess/Mother figure appeared, illustrating the expression of a new quality of spiritual consciousness. The physical evidence reveals a poignant awareness of a Stone Age interest in life's spiritual dimensions. The prehistoric growth of physical, mental, emotional and spiritual intelligences will be considered in more detail later.

Is this 40,000 year continuum of the growth of the human mind an expression of a natural process? Do we have sufficient knowledge of this process to consider that this possibility exists? The idea of our past providing the foundation for the future is mirrored by countless examples from nature where the death of the old supports the rebirth of the young. The gardener's compost heap is one of the best resources for future fragrant flowers and delicious organic vegetables. We see the past creating the future in the sacred geometry of nature. One simple mathematical law at the very heart of Creation is inherent in the many patterns created in the natural world. It is the famous list of numbers known as the Fibonacci progression

[2] Paul Mellars, *Prehistoric Europe*, p.49-50.

where the first and second numbers of a series are added together to create a third.

To create the next number in the series we add together the previous two successive numbers.

When we add the third and fourth numbers, we obtain the fifth: and so on.

1+1=2, 1+2= 3, 2+3=5, 3+5= 5+8=13, 8+13=21, etc.

The simplest examples of this are the following numbers:
1, 2, 3, 5, 8, 13, 21, 34, 55, 89, 233, etc.

The interesting feature of this series appears when we divide any number by the previous number as follows.

3/2=1.5,   5/3=1.67,   8/5=1.6,   13/8=1.625,   21/13=1.615,

In this series we are calculating the ratio of the larger number as a proportion of the smaller number. The first division give the answer of 1.5. As the numbers in this series increase the answer moves towards a perfect irrational state identified as the number 1.6180337..... This number is called the Golden Proportion or *phi* from the Greek alphabet.

These ratios oscillate between smaller and greater values of *phi*, gradually approaching but never reaching exactitude. This rhythmic alternation continues ad infinitum towards *phi*.

The mathematical characteristics of various Fibonacci Series are ubiquitously displayed in natural phenomena. 'The Golden Mean Spiral', in which the geometric increase of the radius is equal to *phi* is found in nature in the beautiful conch shell *Nautilus pompilius* which the dancing Shiva of the Hindu myth holds in one of his hands as one of the instruments through which he initiates creation'.[3] This golden ratio is a fundamental feature expressed in nature. The branches on the stems of plants and the veins in leaves are arranged according to this ratio. This proportion is contained in chemical compounds and the geometry of crystals. It is commonly recognised in paintings and music that capture the aesthetic of beauty. The study of the way the octaves of musical harmonies follow the law of mathematical proportions established the study of the exact science contained in music.

This example of scales of vibration illustrates one of the countless expressions of the principle (rather than the law) of correspondences. The ancient wisdom teaches of its existence throughout life's spiritual, mental, emotional and physical dimensions. This arises from the unity of all things and their mutual sympathy; 'as above so below'[4], as below so above; as within so without. This principle embraces all aspects of life and exercises perception to scroll through its many dimensions and recognize the resonances that metaphor, analogies and simile depend. It creates a chain of understanding between life's dimensions of reality.

A physical example from 1863 science concerns the English chemist John Newland's discovery that patterns are found in chemistry. Newlands classified the 56 then known elements into 11 groups identified by their similarities. He proposed that the law of octaves existed in the relationship between elements based on a multiple of eight in atomic weight. In 1869 Mendeleev, the father of the periodic table of elements published his work on *'The Relationship of the Properties of the Elements to their Atomic Weights'*. He identified a pattern in the properties of many of the known weights, which enabled him to predict the existence and properties of elements later discovered.

---

[3]   Robert Lawlor. *Sacred Geometry*, p. 56.

[4]   *The Kybalion.*

Such atomic structures of materials represent a scale to which fractal patterns move. 'Scientists have shown how fractals, mathematical designs that repeat their patterns on infinitely smaller scales exist in many structures in nature; from bolts of lightning to cauliflower heads, fractal patterns exist. Fractal solutions in the form of equations that create fractal patterns can help predict events based on natural cycles'.[5]

This overview of intelligence as a natural phenomenon found in all aspects of life does not surprise us. However, an understanding that the emergence of the intellect as an addition to human intelligence, as part of a natural process that is expanding physical, emotional, mental and spiritual dimensions of consciousness, is only vaguely recognised. The fact that each generation builds its world on the legacy of its forbears suggests that the growth of our intelligence is a natural multi-dimensional phenomenon. In fact, we know this is the case as each child builds the foundation for its own life from the physical, emotional, mental and spiritual resources of their family. If the principle of correspondences applies to the growth of our own species, the way an infant develops into an adult provides us with a pattern assessed by psychologists to provide some useful ideas about the way the human race has advanced by increments during past millennia.

To see how these ideas work we will first identify some of the basic dimensions of an infant's growth. Observing the behaviour of a young child reveals how after their initial feeding and sleeping period the curiosity of its young consciousness brings countless objects to its mouth. Gradually the skills learnt coordinate the use of different muscles so that the child is able to make the physical movements needed to explore their immediate environment. Learning to walk extends their ability to explore whatever attracts them. The games played extend

---

[5] Shapir and Jorne, *Nature's Cycles in a Fractal State of Mind,* University of Rochester.

this physical intelligence to include emotional lessons whereby unpleasant experiences may teach the need for caution or retreat. After some basic abilities are mastered, social skills beyond those lessons taught within the immediate family develop. The emotional likes and dislikes of young children influence how they play and argue with each other. It is with their first friends they swap toys and treats and obtain an intuitive intelligence about what they want, what is fair or feels right. At the mental level, the lessons from school begin the process of training the mind within the different subjects of the curriculum. Eventually, wanting to make models or dresses introduces new tools to use. Screwdrivers and scissors must be handled with proficiency if success is to be experienced. All these stages of childhood are commonly watched over by the infant's original deities – Mum and Dad.

The growth of all species, whereby they mature and create their own offspring and contribute to the ecology of their environment, express the archetypal energies appropriate for their species. When we look at the evidence archaeologists have discovered to understand human development during past millennia, the archetypal energies expressed by humans are not fixed: qualities of consciousness are still evolving. When we look at how the conscious intellect has contributed to our mind's growing intelligence during recent millennia, we find that new skills always depended on making new physical, emotional, mental and spiritual connections and remembering the useful ones. Historical evidence of this process arises whenever a culture adopted new ideas that changed traditional ways of life - when a consciousness emerged that was different from that of their ancestors.

To identify a second prehistoric era when the intelligence of our ancestors had huge opportunities to develop, we need to recognise when our planet changed dramatically enough to revolutionise the way humans lived; when they began to create cultures from which our own have developed. One such time was when the warming and increasingly wet climate ended the last Ice Age. The dramatic environmental changes created the conditions in which the Neolithic Revolution, c. 11,000 BC took place, some 26,000 years after the c. 37,000 BC Palaeolithic Revolution.

During the millennia between these revolutions, the last Ice Age, which in total lasted about 120,000 years, ended. The resources needed to sustain Stone Age life during this time waxed and waned according to those naturally available in the changing environment. Human existence usually depended on finding a cave or rock shelter that was close to sources of food and fuel. The coldest time of the last Ice Age, the last glacial maximum c. 18,000 BC, was followed by a gradual rise in temperature. It is estimated that it took some 11,000 years for the sea level to rise some 120 metres to the current level and for the world's different climates to stabilise. Before the new equilibrium was established, the melting ice caused many huge floods that wreaked havoc across the world; all low-lying coastal regions were inundated. Many places far from the sea experienced turmoil too. Examples of major events include the ice melting from the Russian Ural Mountains and flowing through the Aral, Caspian, Black and Mediterranean Seas to reach the Atlantic Ocean. In the North American West, the Scablands illustrate the erosion caused when a vast ice dam broke releasing a huge wall of water that ploughed across several states into the Pacific Ocean. In Europe, the British Isles became islands around 6000 BC after floods finally established the boundaries of the English Channel and the North and Irish Seas. Well away from the melting ice, the Gulf of Persia was flooded due to the rising sea level. The depth contours of this gulf suggest that its shallows were flooded around 6000 BC as well. Another natural consequence of the melting ice is graphically illustrated by the changes that took place in Scandinavia. A mountain in northern Sweden provides evidence of the Pleistocene coastline over 280 metres above the current sea level.[6] This isostatic recovery allowed new lands in the northern hemisphere to rise out of the sea to provide opportunities for people to explore and settle.

Living in low lying coastal regions with the sea level continually rising over several millennia, we would expect cultures to have established traditions directly related to the threat of flooding. Part of their security possibly came from identifying new locations for their community in the event of a sudden rise in the sea level.

Personal safety cultivated by identifying escape routes of likely threats.

Archaeologists and historians have identified many of the challenges that influenced the emergence of the new expressions of intelligence. We will begin by considering how these shaped cultures during the millennia since the last Ice Age. This research provides ample evidence for us to assess if changes in the human mind, and their associated civilizing influences on cultures, are part of a natural process.

It will help to clarify how natural processes may be seen in very different ways. We know that nature is 'red in tooth and claw' and each species has an innate intelligence to survive by following the laws of the jungle. Yet, all these apparently separate life forms seem to be part of Life's greater intelligence. This may have an awareness of the opportunities for the lives of all members of each species to contribute to any purpose Life may have. If this were so, each species would likely evolve within this common energetic structure. If this applies to our emerging consciousness, the growth of our intelligence during past millennia may have some clues that help us understand what has been happening.

In the next four chapters, we will begin our detective work to appreciate how intelligence has slowly grown. In Part II, we will consider this evidence to see if it describes a higher intelligence stimulating the expression of human potential.

---

[6]   David Cuff and Andrew Goudie, *The Oxford Companion to Global Change*, p 208.

3

Survival Skills and the Growth of Physical Intelligence

When the last Ice Age was ending, rising sea levels flooded the low-lying coastal plains on which many Stone Age communities had lived for millennia. With global warming again melting glaciers and polar ice fields, the rising sea level threatens the homes and business of millions of people. The rise in carbon dioxide levels from burning fossil fuels contributes to this menace. Our control of industrial practices is not reducing this pollution. Continuing global degradation from other human behaviours is increasing the dangers we face. Industrial agriculture, the over-fishing of the oceans and the destruction of the rain forests are obvious examples of the need to ensure environmental improvements are managed so that long-term sustainability is achieved.

While political lip service recognises these global challenges, the pace of addressing them fails to recognise the threats that influence the survival of our planet's capacity to support the lives of future generations. The seemingly inevitable changes looming will likely be a challenge more profound than many of past centuries.

When the consequences of past excesses are paid for, we may learn of a common sense Ken Wilbur writes of in his book *No Boundary*. His insights focus on what we believe our bodies want in order to make us happy. This usually ignores what nurtures and satisfies a more authentic, deeper sense of identity rooted in feelings of being who we are.[1] Cultures survive by people behaving within existing social structures

[1] Ken Wilbur, *No Boundary*, p. 106.

and traditions. This practice would suppress the latent capacity to transcend the artificial limits imposed by the status quo.

When we consider the millennia following the end of the last Ice Age, we can identify a process that includes the growth of various dimensions and qualities of intelligence. The different capabilities these have given humanity seems to be inadequate skills with which to address our future. To find better ways to enhance the expression of intelligence we will first trace how different qualities of intelligence developed.

To trace these changes and explore the motivation that led to the continuum of success humanity has achieved, we now consider the climatic changes our forebears made since the last Ice Age. At the coldest time of this age, the amount of water taken from the sea to create the vast ice fields and glaciers of the northern hemisphere lowered the sea level to some 120 metres below its present level. As the ice gradually melted, the rate at which the sea level rose did not relate directly to the melt rate. Ice dams created many vast inland lakes and seas across northern America and Asia. Estimates for when these dams broke include their cooling effects on the oceans when vast quantities of ice melt reached the seas. NASA, G.I.S.S. has published details of many of these variations.[2] Between 17,600-16,800 BC, the ten-metre rise in sea level was the first of four surges of melt-water that NASA estimates took place during around 12,000 years. During this process, a period between c. 12,700-10,800 BC, the Bølling-Allerød, saw a sudden, substantial rise in temperature. Archaeological and botanical evidence dated to this period of 1700 years reveals how the climate in the Near East then supported the growth of vast quantities of a wide variety of flora and fauna. The greater availability of food fuelled a human population revolution. It is believed that this period ended when ocean temperatures

---

[2] Vivien Gomitz, *The Great Ice Meltdown and Rising Seas*, Goddard Institute for Space Studies, June 2012.

cooled as vast volumes of ice melt reached the seas. The resultant climatic impact changed the warm and moist climate of the Bølling-Allerød into one that was as cold and as dry as at the time of the last glacial maximum. This cold period, the Younger Dryas (mini ice age) lasted for between 1000 and 1,300 years before the temperature abruptly rose by some 7°C to finally end the Ice Age, c. 9600 BC. Estimates for the time taken for this last change vary between a few years to several decades; within a generation, the warmer and wetter climate again transformed the environment and demanded that cultures learn new skills to survive.

The use of this date disguises the incremental steps taken over many millennia as our Stone Age ancestors adapted their traditional ways of life as demanded by changing environmental conditions. We will now consider the results from several regions affected and how new expressions of intelligence emerged.

New Survival Skills in Europe.

The ocean depth contours of modern maps showing the coastal regions flooded as the last Ice Age ended provide a guide to those places to study human activities where floods displaced the inhabitants. On-going investigations include the flooding that took place in northern Europe. Here two very different scenarios unfolded as the Ice Age ended. Although the sea level rose by 120 metres to create the British Isles, changes that are more complex took place in Scandinavia. In Norway and Sweden where the ice was thickest, the land carried the greatest weight of ice. As this load melted away, these lands floated higher on the molten rock under the Earth's crust. The old shoreline gradually rose out of the sea. As this was happening, an ice dam created a rising lake that flooded the lands around the Baltic before it melted.

As the ice frontier of the frozen tundra of northwest Europe receded the herds of migrating reindeer moved northwards with it. The hunting and gathering culture that depended on this source of food had to introduce new practices to survive the threat of floods in the new regions becoming available to them. Whenever floodwater inundated a settlement, we cannot doubt the care people took to help their family and friends find security and stay together. With this threat of personal separation existing for millennia, it would be surprising if survival skills did not develop for the peoples affected. When a flood threatened a family group it is possible floating trees, logs or rafts were used to escape. If such primitive craft existed, their builders had to learn to discriminate between the suitability of different woods and develop the skills and confidence to use them when necessary.

An early history of European boats, published by the University of Edinburgh's Department of Archaeology, edited by Clive Bonsall, was included in the May 1996 edition of, *Mesolithic Miscellany*. In this edition, the paper by Grigority M. Burov focused *on Mesolithic means of Water Transport in northeastern Europe*. This paper assesses the early use of boats.[3]   The 10th-9th millennium BP, [Before Present] radiocarbon years for Danish and northern German wooden paddles indicate they are older than those discovered across northern Europe to the Trans Urals and the boat rock art from the Pontic Caspian region.   The BP scale of time, when converted into calendar years, (BP into BC), translates to c. 9,500 BC by using the IntCal09 calibration curve. People seated on floating trees, logs or rafts may have been the first to use paddles. Evidence of actual boats of this age does not exist. Burov adds that decorated paddles may have had some ceremonial purpose. In China, a canoe and paddles (c. 8000-7300 BP) were discovered at the site of Kuahuqiau located on

---

[3] Dept. of Archaeology and Mediaeval History, University of Simferopol, Ukraine.

the Lower Yangzi River in Zhejiang province near Hangzhou. [4]

The Near East

Another region where survival demanded a new culture was inland from the eastern Mediterranean, including the valley of the river Jordan, the Levant. Sea depth contours suggest minimal coastal flooding from the sea as the Ice Age was ending. The warmer and wetter climate introduced a wide range of opportunities and challenges for hunter-gatherers. To understand how these changes influenced cultural developments in the Levant as the Ice Age was ending we will refer to the article written by Ofer Bar-Yosef, Professor of Prehistoric Archaeology at Harvard University. The article titled *The Natufian Culture in the Levant; Threshold to the Origins of Agriculture* describes the changes that followed c. 14,500 BP (c. 12,700 BC). The cold and dry Ice Age climate changed to become wetter and warmer. During the period c. 13,000 – c. 10,000 BP the Natufian Culture spread across the Levant with settlements and sedentism playing a major part in the changes that introduced agriculture.

In his book *After the Ice – A Global Human History, 20,000-5000 BC,* Prof. Steven Mithen, Professor of Early Prehistory at Reading University, England provides us with additional information about environmental and cultural changes that took place in the Levant. In the chapters about Western Asia Prof. Mithen writes of the period that began around c. 12,700 BC and lasted for nearly 2000 years when a sudden rise in temperature created a climate that was able to support a profusion of game to hunt and wild foods for gatherers to enjoy. This period is termed the Bølling-Allerød and lasted until around 10,800 BC during which time the new Natufian culture of sedentary hunter-gatherers lived in villages built in the Mediterranean hills alongside the Jordan Valley. Artefacts

---

[4] http://www.antiquity.ac.uk/Projgall/liu/

discovered in this region included stone pestles, mortars together with flint cores, and flakes. Many of the flint blades were found to have dried films of sap from cutting plant stems. This evidence suggests that the microlith blades were from sickles used to harvest crops, possibly wild barley or wheat [5].

After these warm millennia, the climate changed and introduced the mini Ice Age termed the Younger Dryas. Evidence from excavations at the Syrian site of Abu Hureyra, dated to the start of this period reveals a considerable interest in edible grains. This site, formerly on the banks of the Euphrates, before being submerged by Lake Assad, provides extensive evidence of a developing agricultural economy. Discoveries at this site have identified over **150** different species of edible wild seed dated between 11,090 BP and 10,250 BP.[6] These Carbon 14 dates when calibrated into calendar years give dates of c.11,000 BC and 10,200 BC.[7] In addition, it is likely that roots and plant leaves were also staple foods of the time.[8] The botanical interests of people from the site of Abu Hureyra may have been some possible help to the c. 10,000 BC inhabitants of Hallan Cemi and other villages in the foothills of the Taurus Mountains in south-eastern Turkey. This region had pockets with warm microclimates unusual during the Younger Dryas mini Ice Age from 10,800 BC to 9.600 BC. Dr. Michael Rosenberg, from the University of Delaware, headed the team excavating at Hallan Cemi on the Sason River a tributary of the Tigris.[9] 'They found the remains of several circular stone houses and dozens of tools fashioned out of fish and animal bones, stone mortars for grinding nuts and other

[5] Prof. Steven Mithen, *After the Ice*, p. 29-30.

[6] Gordon C. Hillman, Susan M. Colledge and David R. Harris, *Foraging and Farming*, p. 245 and 266.

[7] P. J. Reimer et al, *IntCal09 Supplemental Data - Radiocarbon*, p. 1120

[8] Gordon C. Hillman, Susan M. Colledge and David R. Harris, *Foraging and Farming*, p. 265.

[9] Interest in this site is recorded in 1994 when The New York Times wrote about this work.

food. A surprising discovery was the bones of domesticated pigs that may have been the first animals raised for food'. The findings from three other Near East sites, in use during the early Holocene, illustrate the possibility of some guiding sages behind the changes identified.

The site we will first consider is the southeast Turkish hilltop site of Göbekli Tepe. It was built c. 9000 BC, overlooking the Euphrates near the ancient city of Urfa some 200 kilometres to the north of Syrian site of Abu Hureyra. Near to Göbekli, the wild wheat has the same genetic strain of modern domestic wheat.[10] The Smithsonian Magazine of November 2008 describes how the German archaeologist Klaus Schmidt discovered what he believes to be the world's oldest temple. Its twenty-two-acre hilltop site with breath-taking views across the surrounding countryside is recognised as being of unprecedented archaeological significance. Early excavations revealed five circles of large carved stone pillars, with a central pair of T-shaped pillars weighing between seven and ten tons, the tallest being sixteen feet high. Much larger columns have also been discovered on the site. The Göbekli constructions represent a huge cooperative investment over many generations and suggest a sacred social priority shared by leaders from the surrounding region. Some of the stone pillars have elaborate reliefs that Schmidt considers depicts stylised humans; others have symbolised foxes, lions, scorpions and raptors. These creatures, together with sculptures at the sites of Nevali Cori and the later ones from Catalhoyuk could be symbols associated with some obscure cosmology.[11] If this was a zodiac of the year's seasons, shaman-priests could have led seasonal festivals and temple rituals. These could have included the worship of creatures identified on temple carvings for the protection they afforded against animal and insect pests that

---

[10] George Willcox, *Biodiversity in Agriculture, Chapter 4*, p. 106.
[11]   Ofer Bar-Yosef, *Biodiversity in Agriculture, Chapter 3*, p. 69.

threatened agriculture's birth.

The second place, some 100 kilometres to the south of Göbekli in Northern Syria, is the now flooded site at Jerf el Ahmar, once close to the bank of the upper reaches of the Euphrates River. The findings from this and other c. 9000 BC sites nearby support the idea of a pre-domesticated grain economy. The stone buildings excavated at Jerf el Ahmar discovered rooms for pre-domesticated grain storage and others with three saddle querns side by side for grain grinding. The practice of growing new staple foods could have been a sacred cultural practice organised from Jerf el Ahmar with rituals for everyone to help with the distribution and sharing of seeds.

The third region includes several sites along the Jordan valley. The most famous is early Neolithic Jericho (10,500-9,300 BC), where the early inhabitants farmed domesticated barley and emmer wheat.[12] In this valley, hunter/gatherers planted seeds and used its spring waters to grow the new staple foods that supplemented their traditional supplies of meat. With effort invested in places where seeds were sown and tended, a farming culture was established as settlements spread. This new regional practice of growing supplies of new staple foods introduced the Neolithic Revolution.

To the north of this valley, the land was rain-watered and extended through the Levant, around the foothills of both the Taurus Mountains of southern Anatolia and those to the west of the Assyrian plateau and Zagros Mountains. The major rivers of this region, the Euphrates and the Tigris, defined this region's vast strategic resource, the fertile alluvial plain of northern Mesopotamia. Practical physical intelligence was used to farm this land between the rivers. The changes from hunting and gathering to relying on agriculture and living in permanent villages with stone buildings took around 2,000 years and could

---

[12] Ofer Bar-Yosef and Mordechai E. Kislev, *Foraging and Farming, Chapter 40*, p. 636.

not be reversed.

With the return of a warming climate around 9,800 BC, the region extending from the Jordan valley, through the Levant and around the foothills of the mountains of northern Mesopotamia, known as the Fertile Crescent, its hunter-gatherers were the first to begin supplementing their traditional skills by learning the skills needed to grow cereals. This cultural change to farming introduced the Neolithic Revolution.

In other regions of the world, expanding populations were motivated to introduce their own Neolithic Revolution when hunting and gathering were unable to sustain growing communities. These examples of knowledge illustrating physical intelligence have attracted botanical and archaeological research wherever this cultural change to grow food took place.

It is likely that by around 8,000 BC agricultural experience with many wild plants preceded the selection of the species domesticated by repetitive planting of non-shattering seeds grown as staple cereals. Plants were domesticated when they did not shatter and scatter mature seeds on the ground. When ripe the domesticated plant retains its seeds, a convenience for harvesting. The creation of domesticated crops is interpreted as purifying the species to make their farming more practical. Each domesticated crop required the farmer to plant the seed. The time taken for this first change to take place varies. Some scholars argue that for some species the process can take less than two centuries while others maintain it takes some 1,500 to 2,000 years.[13] [14] In addition to the choice of seeds cultivated, many other factors influenced agricultural success. The suitability of the soil and the availability of water were as

---

[13] George Wilcox, *Biodiversity in Agriculture, Chapter 4*, p. 102.
[14] Dorian Q Fuller, *Biodiversity in Agriculture, Chapter 5*, p. 113.

important as gaining an understanding of the influences on farming of the annual rhythms of the seasons.

The growing of crops enabled communities to secure larger and more reliable sources of staple foods. Farming knowledge changed the balance of power between humans and their environment. A fundamental change took place as *reactive* cultures became *proactive*, a co-creative relationship developed between people and their environment. This rebirth depended on individuals using the potential of specific plants able to provide new staple foods. Cultures formerly dependent on hunting and gathering were transformed as they learned about the relationships between the annual cycle of the seasons and the growing of crops. As farming skills spread, they increasingly sustained larger populations.

A detail that served the harvesting of crops was the use of razor sharp obsidian blades made from volcanic glass from the Taurus Mountains of southern Anatolia. This rich source of raw material for making scythes, centred on the settlement of Catalhoyuk in the Konya plain, is one of the most remarkable Neolithic sites from the 9000 -7000 BC period. The theme of the limited availability of a valued resource possibly created one precedent for the later trading culture that developed. To understand more about the archaeological excavations of part of this huge 32-acre site we will first refer to Prof. Mithen's study of the living accommodation discovered at Catalhoyuk, 9,000-7,000 BC. It was comprised of many individual mud brick buildings constructed against each other but without any ground entrances. Ladders to pathways across the rooftops led to trap doors to enter the rooms below. In the dark confines of this maze of rooms, many frightening images including bull's skulls and vultures record a culture fearful of the natural world. Does this architecture suggest that safety was sought by crowding together? The idea of non-separation from family and friends echoes the security sought in distant lands threatened by floods. The theme of keeping loved ones close, including the deceased, later extended across Mesopotamia by burying the dead under the floor and rebuilding rooms on top of earlier dwellings. This ritual developed into the later practice around 7500-6300 BC of preserving and displaying decorated skulls as described by Prof. Mithen in *Chapter 10, The Town of Ghosts*.[15] This custom may relate to a culture that revered psychic contact with beloved ancestors.

By c. 7000 BC, farming crops provided secure food supplies for the growing tribes and neighbouring communities to expand and prosper by supporting a growing trading culture. The necessary revolution in human survival gradually shaped cultures in ways that provided the foundations for more types of physical intelligence to emerge.

Another region requiring new survival practices during the Early Neolithic was Egypt's inhospitable Western Desert. After the end of the last Ice Age, for about 6,000 years this region had some summer rains that supported a scant savannah with limited opportunities for seasonal hunting and gathering. After c. 8000 BC, Early Neolithic activities in this region is identified from bones and ceramics discovered at sites around Nabta Playa and Bir Kiseiba. This evidence suggests that cattle were first domesticated there to supply milk and blood, rarely meat for food. The discovery of wells for the watering of animals date from c. 7500 BC.[16] As the Sahara Desert expanded eastwards, the herding of domesticated cattle contributed to the agriculture of the Nile valley.

The Growth of Physical Intelligence

Within a few months of birth, young babies move about learning to explore their immediate environment by making

---

[15] Prof. Steven Mithen, *After the Ice*, p.81-82.
[16] *The Oxford History of Ancient Egypt, Prehistory* by Hendrickx and Vermeersh, p.28-9.

physical connections. Gradually the skills develop to coordinate different muscles so that the child is able to play and walk. After the abilities to gain control of the physical body have been mastered an introduction to the tools of the adult world start to be of interest. Screwdrivers and scissors must make precise movements if success is to be tasted. Learning to ride a bike is often a milestone on the way to the teenager driving a car.

During the last fifty years, the flood of knowledge that has become available from the use of computers has created a modern culture dependent on information technology. Learning about IT has become a modern survival skill for living in the global village. The physical intelligence needed to write with a fountain pen is likely to become as obsolete as being able to sharpen a quill.

Cultural revolutions have long been associated with the changes needed to cope when living in new environments. Early evidence of this comes from the changing environment created as the last Ice Age was ending. The threat of floods probably led to the invention of boats. The use of floating logs likely served Mesolithic hunter/gatherers when the threat of drowning arose. From such beginnings, rafts and boats were necessary for their culture to develop. While the warming climate of the newly populated northern lands demanded the use of boats, their success required the necessary knowledge to select the most suitable timber to build practical designs. A boat building culture developed alongside the skills to row or sail them safely before they both became a cultural tradition. It is probable that it was only after the Mesolithic cultures of northern Europe mastered seamanship that they fully benefited from the vast resources of the virgin territories released from the ice. The first evidence of boats confirms their use some millennia later. In the publication *Mesolithic Miscellany*, Grigoriy M. Burov describes late Mesolithic developments in boat design as follows 'Well preserved dugout canoes radiocarbon dated to the

6th millennium BP were revealed in Demark (Tybrind Vig, Prastelyngen – Andersen 1985: Smith 1982: 141-142)'.[17]

In northern Europe, the vast natural resources created as the ice melted were able to support a growing population for several thousand years. In an overview of the European Mesolithic, Prof. Mithen describes how the archaeological record of this period provides comprehensive evidence of the great capability expressed at this time for humans to live in harmony with their natural environment.[18] It was not until this region's natural larder became too small for the growing population that a Neolithic Revolution introduced agriculture to produce the needed foods. In the Near East, another sphere of knowledge that illustrates physical intelligence being directed towards a growing specialised activity of the period is the advances in metallurgy. This subject began to attract human attention around the time of the Neolithic Revolution in a region close to Göbekli Tepe. The article *Development of metallurgy in Eurasia* [19] includes an overview of the use of copper. This began in eastern Turkey and northern Iran with the desire to adorn the body before and after death with coloured ores. Different coloured copper ores used for 'beads, pendants and pigments' were made c. 10,000 BC. In eastern Turkey, evidence of the use of heat to anneal and soften naturally occurring copper dates to c. 8000 BC. The production of metals from metal bearing ores introduced metallurgy. These processes provide evidence of the expression of physical transformations of rocks. Ores were destroyed to be reborn as pure metals. Objects made from these new materials became highly valued within prospering agricultural economies.

Across the Fertile Crescent the cultivation of grain crops, the

---

[17] Dept. of Archaeology & Mediaeval History, University of Simferopol', Ukraine. *Mesolithic Miscellany* V.17, N.1, p.12.

[18] *Prehistoric Europe*, edited by Barry Cunliffe p. 80 -81.

[19] Roberts, Thornton and Pigott, *Development of metallurgy in Eurasis.* Antiquity 83 (2009), p. 1012-1022.

herding of sheep and goats initially supplemented, and later largely replaced hunting and gathering. Local populations flourished as the new skills farming demanded provided a new strand of physical intelligence that hunter-gatherers introduced into their culture. The farming culture successfully practiced at Jericho spread across the northern plains of Mesopotamia. Although the fertility of the Jordon Valley became exhausted, it provided the model that gradually flowed down stream with the Tigris and Euphrates to nurture the birth of the city.

A vital step in this process is revealed from archaeological evidence c. 7,500 BC, which describes a new cultural diversity arising from the physical intelligence to move agricultural produce to places where needed. This introduced a trading culture that included the activities adopted as hunter-gatherers exchanged meat for seeds and flour. From northern Mesopotamia, some 100 km southwest of the site of ancient Nineveh (close to modern Mosul) and south of the Sinjar Plain, the site of Umm Dabaghiyah is known as a centre for its early pottery. It was also a hunting base used to process products to trade from the carcasses of onagers (a wild breed of ancient small horse) hunted on the plains.[20] At the nearby small village of Tell Sotto, the grave of one wealthy farmer contained crafted stone and metal objects from distant places.[21]

Sites for seasonal hunting camps from the Assyrian foothills became farming settlements where the rainfall was sufficient. Estimates for the date and extent of later trading activities come from the distribution of pottery produced from established settlements with all the paraphernalia necessary for economic success. Prof. Mithen describes the site of Yarim Tepe c. 6500 BC as a vibrant centre with its different craft skills including evidence of copper jewellery.[22]

---

[20]   H.W.F Saggs, *Peoples of the Past – Babaylonia*, p. 24.
[21]   Prof. Mithen, *After the Ice* p. 436 .
[22]   Prof. Mithen, *After the Ice* p. 437.

Hassuna pottery from a site closer to Mosel dates from around 6000 BC. Further to the south, pottery from the Samarran culture dates from around 5500 BC. This region of Mesopotamia does not have adequate rainfall for agriculture but provides early evidence of the use of irrigation canals.[23] It became a resource when physical intelligence was behind the idea to move large volumes of water to where it could be used. Land became available to farm that was otherwise unsuitable. In the millennia that followed, the knowledge necessary to manage the successful hydraulic economy led to a culture dominated by the City States of Southern Mesopotamia.

South Eastern Europe.

The physical intelligence from the Near East influenced changes taking place further north. 'Archaeology from c. 6700-6500 BC tells us when the first pioneers moved from Anatolia to colonise Greece'.[24] The evidence of new staple foods being grown in adjoining localities reveals the spread of agriculture. This change moved in two directions. The rich deposit of human occupations from the Franchthi cave in the Greek Peloponnese provides evidence of this new culture beginning to spread around the Mediterranean coast.[25]   Agriculture also moved northwest along the Danube valley c. 6000 BC providing a second route for the spread of agriculture and metalworking skills. The spread of pottery with distinct linear patterns is used to trace this physical distribution. 'In a brief spurt of perhaps 700 years in the fifth millennium, it crossed the 1500 miles between present day Romania and the Netherlands.[26]   In *Prehistoric Europe* Andrew Sharratt writes, 'The later Neolithic and Copper Ages, therefore, represent one of the most complex and interesting phases of European development, during which

---

[23]   H.W.F Saggs, Peoples of the Past – Babaylonia, p. 24.
[24]   David Anthony, The Horse, the Wheel, and Language, p. 77.
[25]   Alasdair Whittle, *Prehistoric Europe*, p. 137.
[26]   Norman Davis, *Europe a History*, p. 74.

the implications of the first spread of agriculture worked themselves out, to be rapidly followed by a second generation of agricultural and livestock-rearing innovations that followed hard on their heels'.[27]

The emerging physical intelligence identified in this chapter reflects how the priority of survival initiated an expansion of new physical skills needed for farming that flowered as the spread of Neolithic Revolution reflected the growing population's intent to prosper.

---

[27] Andrew Sharratt, *Prehistoric Europe*, p. 172.

4

The Growth of Emotional Intelligence

After the changes created by the ending of the last Ice Age, many communities had to learn new skills to survive. This created the new types of physical intelligence that became the foundation for future cultural knowledge. In the Near East, this included the making of the necessary connections for the movement of traded goods. We will now look at how intuition, as one dimension of emotional intelligence, enhanced the growth of physical intelligence.

A dictionary definition of intuition describes 'the power of the mind by which it immediately perceives the truth of things without reasoning or analysis.' Prof. H. Wildon Carr defines intuition as 'the appreciation by the mind of reality directly as it is, and not under the form of a perception or a conception, nor as an idea or object of the reason, all of which by contrast are intellectual apprehension'.[1] In this work, intuition is one of four aspects of emotional intelligence.

The academic study of emotional intelligence accelerated after the two psychologists John Mayer and Peter Salovey published their seminal article on this subject in 1990. They introduced a definition of emotional intelligence by describing 'the subset of social intelligence that involves the ability to monitor one's own and others' feelings and emotions, to discriminate among them and to use this information to guide one's thinking and actions'.[2] These researchers proposed that a model for emotional intelligence includes four different

1 Carr, H. Wilden, *Philosophy of Change*, p. 21.

2 Imagination, Cognition & Personality. Vol 9131 185-211, 1989-90. Baywood Publishing Co. Inc.

characteristics: perception of emotions, thinking about and using emotions, understanding what emotions may mean and managing emotions to provide suitable responses.

In the introduction to the tenth anniversary, reissue edition of the number one bestseller *Emotional Intelligence* (1996), Daniel Golman writes about three models of this dimension of intelligence. The first of Salovey and Mayer is in the tradition of intelligence based on the first work on IQ in the early 1900's. The second model put forth by Reuven Baron is based on his research on well-being whilst the third, Golman's own, concerns emotional intelligence with regard to leadership and high performers at work. These aspects of intelligence are increasingly recognised as dimensions of the psyche able to enhance many aspects of modern life.

During the decade when much progress to understand emotional intelligence was made, scientists at the Institute HeartMath, Boulder Creek, California studying the heart, found it capable of giving us far more than anyone ever suspected. 'It's intelligent and powerful, and we believe that it holds the promise of the next level of human development and for the survival of the world'.[3] It can be seen to provide the capacity to bring wisdom into the emotional life often culturally derided when the intellectual skills of reason and logic work with facts.

The above efforts to increase an awareness of an often-dormant human potential, represents a modern initiative to increase our species emotional intelligence. Clearly, the existing norms, which establish each generation's level of emotional intelligence, can be improved. This type of intelligence describes the intuitive process whereby children learn about their culture's different spectrums of acceptable emotional behaviour. In infancy, whilst the cry of a new-born child brings smiles of happiness, it is possible that after subsequent sleepless months this same sound is less welcome. For many

[3]   Doc Childre and Howard Martin, *The HeartMath Solution*, p. 4-5.

the bond between mother and child ensures a welcome inter-dependence that intuitively evolves in ways that meet the changing needs of the infant. However, sometimes the process is not so positive and the spirit of some children is broken when obeying parental demands. Later some toddlers may learn that expressing the feeling of wanting something by crying is not as successful as it once was. Behaviour, like being nice or good, may teach a particular expression of emotional intelligence. Childhood learning opportunities are emotionally successful when achievements are rewarded with a smile of approval. When the years at school begin, all children come face to face with the standards of behaviour set by the school. These are laid down by the culture to which they belong, and are part of the 'necessary' conditioning for all 'good' citizens - to live in ways that do not break the law, written or unwritten; the experience of approval is the reward. At whatever stage fear of disapproval is able to dictate behaviour, there is a loss of confidence to follow personal intuition – an intuitive knowledge of what is right is not trusted.

To see how different expressions of emotional intelligence built on past beliefs as well as the new physical intelligence considered in Chapter 3 we will look at the changing cultures of south-eastern Europe, the Pontic Caspian region, Mesopotamia and Egypt.

South Eastern Europe.

After the Neolithic Period c. 10,000-6000 BC, the Chalcolithic Period 6000-3000 BC spanned the millennia between Stone and Bronze Ages. This period saw the use of copper for ornaments and the high quality of these grave goods reveals an understanding of metallurgy, whereby copper ore was mined and purified. This metal was then crafted into beautiful artefacts displaying an aesthetic culture valuing excellence. The qualities of copper cannot replicate the sharp efficiency of stone tools.

The paper on the *Development of metallurgy in Eurasia* [4] describes the progress after the Neolithic Period. By c. 6000 BC, the technology for smelting copper and lead to refine these metals had spread to south-eastern Europe and Afghanistan. The modern archaeological trend is to consider how the advances in ancient metallurgy resulted from social and cultural pressures rather than technological capabilities.

These developments illustrate a motivation to obtain the pure metals desired by traditional customs. It is possible skilled 'geologists' travelled extensively to assess the quality of the different types of ore available. Success depended on aware and intuitive, highly skilled specialists learning and teaching the specific technologies and introducing the equipment necessary for processing the different ores. New industries built close to large reserves of high-grade ore helped the communities with these resources at hand to prosper. Successful traders became the new cultural elite with the wealth to make this technology profitable. The use of metal to replace stone and bone for many tools and weapons was only practical after c. 5000 BC when molten tin and copper were mixed to produce bronze. Supplies of tin introduced a new dimension to international trading.

The Pontic Caspian Region.

David Anthony in his book *The Horse the Wheel and Language* provide insights into how consciousness in this region was changing around 5200-5000 BC. In his chapter titled *Cows, Copper and Chiefs,* he describes how, when cows and copper became part of the culture of tribes of the Pontic Caspian the regions' chiefs came into prominence.[5] This

---

[4]   Roberts, Thornton and Pigott, *Development of metallurgy in Eurasis.* Antiquity 83 (2009), p. 1012-1022.

[5]   David Anthony, *The Horse, the Wheel, and Language*, p. 160.

identification of social rank is further evidence of how individuals were/began to be recognised for their personal skill. Anthony describes how different words of the Proto-Indo-European vocabulary identify different positions of high social rank: 'village chief', 'he who held power in a group', 'a regulator', 'one who makes things right'. The spread of farming across the steppes to the north of the Black Sea introduced a new culture in which powerful people were an intrinsic part of its social hierarchy. Communities led by traders, owners of livestock and/or land or priests had replaced the traditional hunter/gatherers. Where the owning of resources became an issue, a person would intuitively realise the power and social repercussion of ownership. Possession and power within a culture became an emotional mixture able to be exercised for good or ill.

The growth of trade between the different cultures identified by Anthony required languages to develop in order to reflect the new connections made between people with different traditions and owning surplus 'foreign' goods. These exchanges would promote communication and mutual learning and inter dependence.

From around 5000 BC, the growth of physical intelligence in this region complemented the emergence of new expressions of emotional intelligence. Just as boats were earlier invented to survive the threat of floods, on land the domestication of the horse and horse riding were adopted as aids for personal transport when herding animals. The emotional intelligence in terms of equestrian skills concerns the intuitive connection experienced between horse and rider. In Anthony's book, he traces the origins of equestrian skills to the Russian steppe of the Pontic Caspian Basin. This location was found after analysing sites where horse teeth have been discovered that reveal the wear on specific teeth caused by a rider's use of bit and reins to control the animal. Studies on the horse culture of

Northern Kazakhstan provide evidence of it beginning about 3,700-3,500 BC. 'The case for horse management and riding at Botai and Kozhia is based on the presence of bit wear on seven Botai-Tersek horse teeth, P2s, from two different sites'.[6] 'It is likely that Botai-Tersek people acquired the idea of domesticated animal management from their western neighbours, who had been managing domesticated cattle and sheep, and probably horses for a thousand years before 3,770-3,500 BC'.[7]

If a culture of horse riding did indeed begin c. 5,000 BC, it is likely that a natural riding culture that did not require the use of a bit developed earlier. Perhaps the inherent intuitive capacity of shaman, aware of the sensitivity of the horse to any threat, a characteristic needed by game animals, experienced a self-awareness with an intuitive connection, which was trusted by the animal. Such a development would then be a prelude for a relationship between horse and rider based on their mutual enjoyment, (a practice increasingly in vogue as natural riding). The later use of a bit to control a horse was then possibly introduced once the cultural validation of domination and exploitation classified the animal as a resource able to be trained to follow instructions.

In the above scenarios, we have two different conditions. In the first case, self-awareness appreciates that the rider and horse have an intuitive link. In the second case, self-awareness recognises the benefits when the horse is used as a source of power and is trained for a purpose. More than a thousand years could separate when these initiatives were first adopted.

The Near East

In Mesopotamia, the 'land between the rivers', a new expression of emotional intelligence arose in response to the

---

[6] David Anthony, The Horse, the Wheel and Language, p. 220.
[7] David Anthony, The Horse, the Wheel and Language, p. 221.

use of the irrigation necessary to benefit from the fertility of this alluvial plain. Farmers of this region lived in fear of losing their crops to the floods that took place when the courses of the Euphrates and Tigris rivers changed. The threat arose as the settling of sediment created riverbanks higher than the surrounding land. When the rivers were in flood, water breeched these banks; it destroyed crops as the river found its new lower course. This process continually repeated itself.

These fears echoed themes from the distant past, the period when the ending of the last Ice Age flooded coastal regions around the world. To appreciate some of the basic changes that took place to the south of Mesopotamia we note that ocean depth contour lines indicate that the greatest depth of the Persian Gulf is currently 102 metres. Around 15,000 years ago when the sea level was 120 metres lower, the valley beneath the Persian Gulf was inland from the Indian Ocean. As the last Ice Age was ending, during the thousands of years it took for new global climates to become established, there were many periods when floods caused the sea level to rise. During these times, any inhabitants of the lands beneath the Gulf of Persia experienced the sea continually pushing them back towards higher ground. How many times this happened when the sea level rose by 120 metres is unknown. The residents of this region would be likely to believe that the sea was trying to destroy them, that the feminine goddess of the sea, Tiamat was their enemy.

In the Babylonian myth of Creation, saltwater is the primordial element and the source of the Great Goddess Tiamat, present in human form or that of a sea serpent. From her, all beings arose including her consort Apsu, a fusion of sweet water. The earth was a plateau floating on the Apsu, which broke through the earth's surface. Tiamat as the feminine element of the sea gave birth to the primitive chaos of the world to be organised by the intelligence of the gods. The male gods believed that the Great Goddess Tiamat would destroy them:

their greatest of gods Marduk, killed Tiamat, as the Earth Mother goddess, and from her body created humanity to serve the gods of the universe he ordered.[8] An echo of this demise of the feminine comes from southern Mesopotamia records: at the Sumarian city of Lagash, the 2350 BC law code punished women with stoning for the tradition of having two husbands.[9]

Controlling the release of river water to irrigate chosen areas of land required high levels of knowledge and management. The different hierarchies with this power were believed by local cultures to be able to mitigate the disasters caused by flooding. How different communities organised the necessary activities is a matter of debate since little is known about the Ubaid culture of this region. It is likely that the learning of different skills took place to nurture and enhance community activities. For the Czech archaeologist Petr Charvat, it was the last of the true egalitarian societies with self-sufficient households, all equally engaged in subsistence activities.[10] It is probable that the temples of the pre-urban Ubaid culture were the communal buildings for the storage and distribution of agricultural produce as well as the worship of the local god. Temple priests were the likely organisers of secular life before the warlords and kings of the later city-states ruled their own domains. In his book *Oriental Despotism – A Comparative Study of Total Power, (1957)*, the German-American writer and thinker Karl Wittfogel coined the term 'hydraulic civilizations' for cultures dependant on the control of water and the hierarchies that organised this by developing their despotic rule. Paul Kriwaczek writes how this theory is not now widely accepted and that Mesopotamian culture provided opportunities for many powerful groups to stimulate contributions to the needed organising bureaucracy.[11]

---

[8] *New Larousse Encyclopedia of Mythology*, p. 51-54.
[9] Leonard Shlain, *The Alphabet versus the Goddess*, p. 52.
[10] Gwendolyn Leick. *Mesopotamia The Invention of the City*, p.15.
[11] Paul Kriwaczek, Babylon – Mesopotamia and the birth of Civilization, p. 13.

The Ubaidian pre-urban culture extended to southern Mesopotamia and dates between 5500 and 4000 BC. Many villages had temples and their residents grew crops, herded livestock and provided the labour needed to build the necessary irrigation canals. It was believed that the gods controlled the floods and the fate of farmland and farmers. This provided opportunities for religious exploitation by those hungry for power.

To the west of where the Euphrates meets the sea, the earliest constructed shrine discovered revealed offerings left to bring the goodwill of Enki, the god of the sweet water lagoon surrounding the temple. Excavations of the many levels of this temple site revealed the initial construction of a small 'chapel' c. 4,900 BC.[12] These foundations are recognised to be those of the first city in the world, Erudi (Tel Abu-Shahrain). Once the temple priests became local authorities who directed the activities of a community, egalitarian traits began to be lost. 'Henceforth the gods take up residence on earth and live in cities. This philosophy identified that the purpose of humanity was to render service to god and temple.[13]

As food surpluses increasingly became available, trading opportunities expanded together with the need for local management hierarchies. In her book *Religion in Ancient Mesopotamia* Jean Bottero describes how the later cities with the 'major temples' at Nippur, Erudi, Uruk, Nimrud, Niventh and Babylon, followed 'the theocentric cult, the objective of which, following humankind's fundamental purpose and 'vocation' was to provide the gods with all the goods and honours they desired'.[14] This religion seems to have been based on the emotional manipulation of traditional beliefs and the established power of each region's warlord within the

---

[12] Ibid, p.6.

[13] Gwendolyn Leick, Mesopotamia The Invention of the City, p.2.

[14] Jean Bottero, Religion in Ancient Mesopotamia, p. 115.

individual city-states of southern Mesopotamia. Sumer became the first man-made environment designed for the trading of all manner of goods. While Erudi retained its sacred status in Sumarian culture, around 3800 BC, the nearby city Uruk became its cultural centre. It came to be regarded as the mother of all the city-states that spread throughout Mesopotamia and its surrounding regions. The new emotional intelligence expressed in the religions of the city-states of Mesopotamia illustrates the growth of the emotional power of the city.

Egypt

Egyptian emotional intelligence associated with the introduction of cattle brings attention to the cultural importance of the bull. From c. 5000 BC for the next 4000 years, the bull was a symbol valued for illustrating Egypt's fertility, wealth and power.   The archaeological discoveries made between the Nile and the Red Sea, in the now inhospitable Eastern Desert describe an ancient dependence on hunting and herding.[15] Toby Wilkinson writes about the introduction of agriculture and importance of the bull as a valued resource in the Egyptian Badarian period, 5000 – 4000 BC. The owners of large herds of cattle achieved the desired cultural status and success because the Bull had become a symbol for wealth and power.   The burial grounds reserved for the ruling elite included cattle. A measure of the cultural standing of the bull is seen in illustrations in which Pharaohs are depicted wearing the tail of a bull as a symbol of the fertility and power of his position in Egyptian culture. During the rule of Rameses II, the wild bull was still regarded as the fiercest of all beasts and a suitable subject within his temple for the Pharaoh to be depicted hunting.

To see how the bull contributed to Egypt's established religious beliefs we first need to gain some detailed insights

---

[15]   Toby Wilkinson, *Genesis of the Pharaohs* by p. 101

into this complex sphere of Egyptian life. For this, we will refer to Stephen Quirk's book *Ancient Egyptian Religion*. Quirk structures his work into five chapters to give an overview of its different themes. This aids the western mind wishing to comprehend some of the spiritual nuances connecting numerous deities. The five themes considered are as follows:

1. Heavenly Power –The Sun God,
2. Earthly Power – Osiris and Horus,
3. The Preservation of the Universe – Kingship and Cult,
4. Extending Life – Protection of the Body,
5. Surviving Death – Transfiguration.

Of these five themes, the possibility of kingship and cult failing to preserve the universe was minimised by the priority the early kings gave to this task.

When we consider Quirk's interpretation of how kingship and cult were structured in ancient Egypt, we want to see if this new cultural order expressed another new aspect of emotional intelligence. From *Ancient Egyptian Religion* we have descriptions of how, in the beginning creation was ruled over on earth by the sun god. This deity was both creation and creator. The beliefs about the movement of the sun god away from the earth to the heavens, according to texts dated from the reign of Akhenaten, concern the damage done to creation by humanity's fall from divine grace.[16]

This damage could be minimised if the sun god had a representative on earth who 'caused Right to exist'. This may be interpreted as the king doing God's Will. This result required that Right be raised up to the sun god (praise?), 'by the gods Thoth and Inheret'. The king and his priests were charged with the responsibility to be 'he who brings back the distant goddess'.[17]

---

[16] Stephen Quirk, *Ancient Egyptian Religion*, p.31.
[17] Stephen Quirk, *Ancient Egyptian Religion*, p. 70.

These beliefs describe a culture, which through the divine power of the king, its temples and religion took on the responsibility of preserving the universe in the absence of both the sun god and Earth Mother. Their absence possibly relates to the times after human activities introduced farming. It is of interest to note here that the Egyptian strategy to heal creation depended on the use of its temples. Their purpose was central to Egyptian beliefs for the preservation of the universe. They were power stations generating the energy believed necessary to ensure creation's survival.[18] The pharaoh was the high priest and the local priesthood dispensed his word. A social division separated the powerful from those who responded to the word of authority.

This patriarchy did not deride the feminine. The introduction of the goddesses Wadjyt and Nekhet, which represented the cobra and vulture protecting Lower and Upper Egypt respectively, served to protect the king in his task of raising Right to the sun god to preserve the universe. Before this, the eye of the creator created humankind. The Eye of Ra is a symbol of profound significance with many interpretations concerning what the male and female eye may see to initiate creation.[19] In addition the divine eye that existed before creation could be envisaged as a goddess, the loving daughter Hathor (the cow) or her raging *alter ego* Sekhmet (the lioness); as the goddess Nebethetepet, 'mistress of offerings', she shared a place in the cult at Iuna. The different energies of these goddesses may be appreciated as distinct expressions of the archetypal energy of the goddess, rather than distinct individuals. When we consider Hathor, she was a type of deity, sometimes the daughter of the sun god, at others the aggressive Eye of Ra as symbolised as a cobra or lioness, yet also the passive, beautiful, motherly goddess. Hathor, when sent by her father to slaughter rebellious humans, was known as the

---

[18] Ibid p. 70.
[19] Ibid p. 26.

fearsome leonine goddess Sekhmet. After Ra tricked Hathor into stopping her wanton destruction of humanity, she returned as the passive, loving and gentle cow-like Hathor. [20]

The Eye of Ra or Horus is a metaphor for seeing and action. When it leaves a deity, it has to be coaxed back or restored. 'The task of bringing the eye back to Ra, of raising Right to the creator, was the equivalent of healing creation.' [21]

Mesopotamian and Egyptian cultures include different circumstances that reduced the worship of feminine deities or placed them in the service of masculine themes.

The growth of emotional intelligence considered in this chapter includes descriptions of how the greater reliability of food supplies, which resulted from the introduction of agriculture, created changes that enriched cultural diversity. Food surpluses encouraged the expansion of trade and allowed people to follow their intuition and develop the natural aptitudes needed for the growth of the craft skills that expanded trading opportunities. This period describes how the emerging consciousness used survival skills to create cultures built on expressions of physical intelligences. These physical abilities correlate with the way the growth of countless natural species reflect the laws of sacred geometry. An introduction to this subject was considered in Chapter 2 and illustrated by a simple arithmetical progression from the Fibonacci Series. The successful use of this idea, to relate physical and emotional intelligence, needs to be expanded to see if these dimensions of consciousness relate to the growth of mental and spiritual intelligence.

When we do this, we will look for more complex dynamics of consciousness being exercised. Examples of this are seen in the way growing emotional intelligence encouraged physical

---

[20] Carolyn Graves-Brown, *Dancing for Hathor – Women in Ancient Egypt*, p. 169.
[21] Stephen Quirk, *Ancient Egyptian Religion*, p. 32

intelligence to make more movements and recognise different connections appropriate to the new level of emotional intelligence being expressed. This feedback process may function with a natural rhythm that introduces adjustments to keep different dimensions of intelligence in tune with each other. Examples of this include the use of boats and horses. The invention of boats and their skilful use to escape from floodwater and the domestication of the horse to herd animals provided the means for people to extend their knowledge of their wider environments.

5

The Growth of Mental Intelligence   *Education*

Education was needed for the culture of writing to introduce the beginning of recorded history in Mesopotamia and Egypt. After records began to be made and play a part within the cultures of these two regions a new source of clues was left for archaeologists and historians to examine and consider. Their work has helped our understanding of the cultural changes that took place in these two lands during the last half of the fourth millennia BC. The discoveries in and around Mesopotamia c. 3500 BC of small clay tokens of many different shapes were likely to have been easily formed by hand before being marked and dried. Those with holes through them suggest they were tied to identify goods. Some of the markings on these tokens are similar to the later pictograms. Cylinder seals were in use c. 3400 BC. The earliest pictogram writings discovered detailed notes in the Sumerian language from the city of Uruk during the period 3000-2900 BC. After 2600 BC, writings were made with a wedged shaped stylus to make a series of triangular impressions needed for each word of cuneiform script.[1] People had to be educated to read and write.

In Egypt, the earliest evidence of labels with hieroglyphs comes from the Naqada III period, Dynasty 0, c. 3200 BC, from a tomb in a royal cemetery at the town of Abydos. At this time the king's name was written in hieroglyphs symbolising phonetic signs. The first Egyptian kings used hieroglyphs in their temples and burial chambers; these sacred carvings were regarded as God's Words, words that initiated Creation. Rather than recording what happened, hieroglyphs were believed to be

---

[1] H.W.F. Saggs, *Peoples of the Past, BABYLONIANS*, p. 48, 76, 182.

statements of what was being created forever.[2] The supreme titles of kings included images of the falcon. The soaring falcon's mastery of the skies was seen as the earthly symbol for the power of the Sun god Ra, the celestial source of life. It was this this visible portion of the Sun's daily cycle, which had established Egypt's original religious traditions followed by the priests of Heliopolis near to the Giza pyramids. The spiritual symbol of earthly power was the Horus falcon and the falcon's head was always portrayed on the king's human body.

During Dynasty 1, the graphic art of the hieroglyph recorded more complex descriptions of the status of kings.[3] Egyptian art appeared among the first texts at the founding of the Egyptian state c. 3000 BC. [4] The art in question includes the god Seth/Set who several centuries later became the symbol of disorder and destruction, the adversary of life's creativity in the myth of the annual Osiris cycle of the Nile flood. The famous struggle between the brothers Osiris and Set may have been based on historical events.[5] A famous artefact, the Narmer palette, from around the middle of the fourth millennium, provides clues about such discord. The fragment of this palette in the Louvre depicts bearded men setting forth to hunt wearing the bushy tail of an animal. At their head marches their leader holding a staff or totem pole, the top of which carries the image of a falcon.[6] The victory of the falcon god's tribe established the lasting significance of this Egyptian symbolism.

We can gain some insights about inter-tribal relationships by piecing together some of the known details of the different regions of the Nile valley and delta. Records allow us to appreciate the virtual permanence of tribal boundaries. During

[2] T. Wilkinson, *Genesis of the Pharaohs, p. 135.*

[3] K. A Bard, The *Oxford History of Ancient Egypt*, p. 60, 74.

[4] Stephen Quirke, *Ancient Egyptian Religion*, p. 54.

[5] J. Viaud, *New Larousse Encyclopaedia of Mythology: Egyptian Mythology*, p. 20.

[6] Ibid, p. 9.

the fourth millennium BC, there were twenty such territories or nomes in the region of the Nile delta comprising the Lower Kingdom. To the south, the twenty-two nomes in the Nile valley defined the Upper Kingdom. Each Nome had its own animal totem and economic cult centre and was 'symbolised by an emblem on a standard as a banner of territorial identity reaching back to the pre-unification period'.[7] Other tribal emblems portray a bird, an animal or a reptile deity each held sacred within a particular Nome. Each generation intuitively followed its tribal tradition of worshipping their chosen creature. The original purpose behind the creation of such beliefs is unknown. It is possible each tribe's shaman led this part of their culture by experiencing the spirit of their chosen creature on behalf of their tribe. Observations concerning the feelings tribes had for their particular deity were sometimes taken to extremes. 'Popular superstition in later times so grew that every individual of the species of animal in whose body the spirit of the provincial god was incarnated was regarded as sacred by the inhabitants of that province. It was forbidden to eat them, and to kill one was a heinous crime.'[8] 'Tribal disrespect for the sacred creatures of neighbouring nomes 'sometimes gave rise to fratricidal wars'.[9]

The characteristics of any species, defining it as a unique expression of life, can be appreciated as being the embodiment of archetypal principles that together ensure the continuing integrity of each species. In *From Intellect to Intuition* Alice A. Bailey writes 'Let us regard all manifestations of life as spiritual, and so widen the usual meaning of this word to signify the energies and potencies which lie [at the] back of every form in nature and which give to each of them their essential

---

7   Charles Maisels, *Early Civilizations of the World*, p. 34.

8   ibid p.46.

9   ibid p.46.

distinguishing   characteristics   and   qualities'.[10]   This
understanding of the energies, when associated with the
different tribal sacred creatures may be related to how each
tribe worshipped their deities' ability to live in harmony with
its environment.

We know that the actions of the first Kings of Egypt
eventually created the necessary support to unify all the tribes
living alongside the river Nile and created the nation of Egypt.
The use of hieroglyphs helped to establish royal power and the
king's divine authority over arching existing religious
traditions. The use of hieroglyphs for religious texts and public
monuments indicates that mental education was an integral
part of early Egyptian culture. It may have helped solve any
emotional difficulties arising between the traditions of its
different tribes. This may relate to the original tribal
relationships concerning the different tribal totems, territories
or nomes that created Egypt's Upper and Lower Kingdoms.

The Egyptian religion based on the divinity of the
king/pharaoh as the incarnation of the Horus falcon, reflects the
supreme place of the falcon in nature's hierarchy of creatures.
This established a cultural correspondence whereby the
traditional worship of sacred creatures was encompassed
under the wings of the falcon to reduce tribal insecurities. This
hypothesis provides an example of mental intelligence creating
a belief structure able to bring greater national harmony since
diverse tribal beliefs were acknowledged and protected within
a celestial context.

The large number of animals whose head appears on
Egyptian divinities shows the popularity of sacred animals
throughout Egypt during the reigns of all the pharaohs. 'The
notion of 'gods of provinces' and 'gods of towns' formed a

---

10   Alice A Bailey, From *Intellect to Intuition*, p.15-16.

crucial binding medium in Egyptian society'.[11] After 500 years, this progress led to the flowering of the Egyptian Old Kingdom with its pyramid culture, c. 2,600 BC and the blessings that flowed for millennia from its shared security and wealth. A very different history from that of the warring city states that replaced the c. 4000 BC egalitarian Ubaid culture of Mesopotamia.

It is widely recognised that education for the expression of mental intelligence through the introduction of written languages has had profound cultural consequences. Included in these results is the development of psychological traits inherent in the cerebral changes required to communicate with written languages.

In parallel with this process, the decline in the power of feminine values is reflected in the way patriarchy has undermined the cultural status of the goddess. The Californian brain surgeon Leonard Shlain in his book The Alphabet versus the Goddess presents insights about the psychology associated with these changes. His professional experience has enabled him to argue about the cultural implications arising from changes in cerebral processes. His thesis relates the necessary characteristics to establish the long reign of patriarchy and misogyny. These are identified from his wide-ranging studies of cultural changes provided by anthropology, history and religion.

Feminine intuition and spontaneity available from the use of the left-brain shun the laws of grammar needed for all writing demanded by the brain's right hemisphere. This theme of the Mesopotamian 2350 BC Hammurabi law code established a culture of patriarchy. Written law and logical truth displaced a culture that valued the subjectivity of intuitive truth.

Exclusive law reigning over personal truth.

Black & white over gray.

11   Stephen Quirke, *Ancient Egyptian Religion*, p.70.

The later Minoan linear scripts originating from Crete preceded the letters of the Greek alphabet and the widespread use of the Greek language around the Mediterranean coast. In India, the earliest sacred texts of the Vedas date from the period 1500-1000 BC. Oracle bones from China date from c. 1200 BC.

In the next chapter, we will use the Greek language and the later writings of the poets Homer and Hesiod as milestones to denote the times when the frontier of growing intelligence moved from a mental agenda to address the challenge of its use to understand what spiritual intelligence may embrace.

6

The Growth and Use of Spiritual Intelligence

Evidence of human activities during the start of the Upper Palaeolithic Revolution, c. 38,000 BC reveals that our Stone Age ancestors expressed imaginative capacities that illustrate spiritual intelligence. For the first time the teeth of carnivores began to be used for pendants. This artistic development suggests a belief that the power of the hunter is aided by the wearing of prized teeth. The famous c. 30,000-34,000 BC statute found in a Hohlenstein-Stadel cave, of a person with the head of a lion carved from mammoth ivory, illustrates a more sophisticated use of the hunter's imagination.[1] A shaman may have used this ivory carving when teaching hunters to identify with the spirit of the lion. The c. 30,000 BC charcoal drawings of the Chavet cave lions may have served a similar purpose. The prey animals also featured in this cave's drawings may be associated with the visualisation of an abundance of this game.

A later stage of the expression of human spiritual growth was the motivation behind the creation of the 'Venus' figurines discovered widely distributed across Europe by 25,000 BC. This feminine theme is believed to relate to the Great Cosmic Mother, an interpretation of a bas-relief sculpture discovered in a Dordogne cave. Archaeologists recognise this image as the 'Venus of Laussel' and date it to c. 19,000 BC.[2] Any of the red ochre paint remaining on this carving can only be used to obtain a radiocarbon date of the paint; the actual carving could have been venerated for millennia before it was painted.

---

[1] Paul Mellars, Prehistoric Europe, The Upper Palaeolithic Revolution, p.50.

[2] Norman Davis, *Europe a History*. p. 72.

The Earth Mother goddess was accepted as the primary spiritual focus that lasted for some 20,000 years. The change from matrifocal or matriarchal to patriarchal societies was a gradual process. The earliest recorded idea of a masculine God did not arise until a few centuries before 3,000 BC, with the arrival of the Egyptian Kings of this time to unify the tribes of the Nile valley. The concurrent Egyptian invention of hieroglyphs has allowed some of their early spiritual beliefs to be understood by modern scholars. The introduction of the mythology of the sun god Ra as the symbol of the Divine Source of Creation required a representative on Earth. This representative of Ra was the Horus falcon, the master of the Egyptian skies and the divine ancestor of the Pharaohs always depicted in hieroglyphs as a man with the head of a falcon. The first kings of the newly unified land adopted the name of the god Horus to denote their acceptance of regal and divine power.

Clearly, symbols provide a language capable of identifying the changeless within the dynamics of creation. The scientist uses numerous mathematical symbols to define this formal order. In all these, Man is the observer rather than the observed. It would be of interest if there were a symbol able to provide some insights about human life on Earth. The Great Pyramid of Giza has been claimed by scholars to provide this symbolism. However, despite being the subject of exacting intellectual analysis, no scientific proof of this claim is possible. This arises from the philosophical basis of the interpretation.

The book *The Great Pyramid Decoded,* by Peter Lemesurier, reviews the large effort of recent centuries to understand more clearly the secrets encoded in this enigmatic symbol. It examines the archaeological and surveying work undertaken to establish the exact details concerning location, orientation and dimensions of all its external and internal features. The challenge of deciphering and publishing the meanings of this numerical language was latterly undertaken by Adam Rutherford in the 1950s and published in his *Pyramidology.*

Lemesurier's work builds on this.

A basic requirement to arrive at these conclusions concerned the units of measurement that were used to unlock the encoded secrets. The Royal Cubit was defined by the dimensions of the passageways and chambers close to the heart of the pyramid itself.[3] This and other precise units of measurement were applied to the surveying results. The pyramid's external dimensions were seen as a symbol for the Earth's relationship with its cosmic environment. In a similarly profound masterpiece, Fig. 6 illustrates many of the internal details of the passages, shafts and chambers. From the entrance humanity's journey to the different chambers is provided by a numerological commentary interpreted from the passing architecture. This interpretation is of Man's journey in time and his developing relationship with the Earth and cosmos.

Fig. 6.1 Details of some internal features of the Great Pyramid

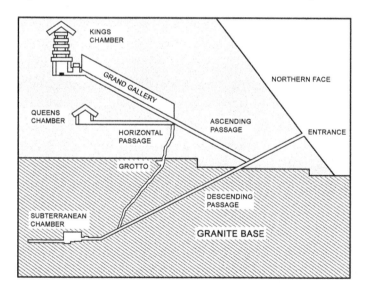

---

[3] Peter Lemesurier, *The Great Pyramid Decoded*, appendix A, p.301

The journey described expresses the potential for humanity to contribute to the dynamic harmony of terrestrial conditions so that we are in balance with the changing cosmic environment as the Earth moves through space. The opportunities to ascend during particular times are tempered with those to descend as well: as dynasties waxed and waned, we have a portrayal of the game of Snakes and Ladders!

The period covered by this forecast extends to the year A.D. 3989. It reads like a commentary on human opportunities to progress, and the consequences of failure. It includes a commentary for the early decades of the third millennium. During this period, the collapse of a materialistic culture is forecast.

The power vested in the succession of Egypt's first kings was accepted as the earthly represetation of the sun god the symbol for the Creator. Some two thousand years later the pharaohs of the 18th Dynasty gradually retrieved power earlier lost to the Amun priesthood. The Pharaoh Akhenaten made them redundant when he introduced a revolutionary new monotheist national religion that worshipped the Sun Disc, the Aten and established the Armarna Period. Much debate surrounds what actually happened during this period including the idea that the Aten was a symbol of Akhenaten's own father Amenhotep III. When he ruled he was the Magnificent King, responsible for bringing Egypt to the peak of her glory, the height of her power, influence and wealth. After Akhenaten's death the powerful Amun priesthood of Thebes tried to destroy all evidence of Akhenaten's reign as Pharaoh. However, popular biblical history records how the monotheist spiritual revolution initiated did not fail. It was accepted in a new guise when the Egyptian Hebrew slaves were inspired through the divine guidance received by Moses to worship their omnipotent monotheist deity. The exodus of Hebrew slaves from an Egyptian captivity is a powerful aspect of the Abrahamic faith mythos. By this time, Hindu religion's Rig-Veda Vedic texts were essentially complete.

The historical origins of other spiritual beliefs concerning monothesim are attributed to the works of many inspired individuals. Those accepting Abraham's teachings recognise his sacred contribution to Jewish, Christian and Muslim faiths. The Torah, the early books of the Old Testament are thought to have been written around 900-700 BC. During the same period the two epic poems, *Iliad* and *Odyssey* are attributed to the legendary Greek writer Homer. Herodotus believed he lived c. 850 BC but other ancient sources suggest an earlier time close to the Trojan War 1194– 1184 BC. Homer's inspired writings introduced a new 'deity', *the spectrum of human experience*. The heroes and heroines involved in the battles of Troy introduced archetypal themes that were accepted into Greek culture's later inclination to interpret their world in terms of archetypal concepts. These primordial changeless absolutes lie behind every dimension of existence, patterned to bring form, meaning and value to the reality hidden by the distraction of appearances. Greek gods and goddesses became the archetypes of human excellence; their extensive pantheon spanned all spheres of human activity.

If we consider Homer as writing around 1000 BC, this is close to 2,000 years after the time the Egyptians introduced a new mental intelligence to teach tribes about the new Sun cult, and its possible contribution towards the unification of tribal lands. Homer's inspired work introduced greater clarity to the idea of heroic archetypes. The protection of sacred animal spirits within the new Egyptian culture illustrates this same archetypal level within the natural context provided by the Nile valley and delta. It was in this environment that sacred creatures were nurtured and recognised as archetypes in life's eternal drama (and therefore subject to the divine rule of the Pharaoh). In composing his *Iliad* and *Odyssey*, Homer establishes the archetypal heroic roots of Greek mythical culture. Its archetypal heroes of Greek mythology were portrayed as being subservient to the wishes of the Greek's pagan gods and goddesses. These Greek themes together with other inspired myths stimulated the widening search for ideas to answer life's wider questions. The work of Thales, c. 636-546 BC introduced a pre Socratic philosophy with the belief that a rational study of nature would reveal life's underlying unity.[4] Other global philosophies were born during the sixth century BC. In China, Confucius c. 551-479 BC, worked at perfecting the art of public behaviour, while in India the Buddha, c. 563-483 BC, discovered the path to end all suffering and introduced the idea of the Buddha of Compassion, Avalokiteshvara.[5]

Hesiod (c. 700 BC) complemented Greek intentions with his classification of his country's pantheon of gods and goddesses. Greek culture blossomed with the works of Socrates (469-399 BC), Plato (427-347 BC) and Aristotle (384-322 BC). Plato's primary philosophical focus was to discover that hidden within the world's emperical changes the eternal truth of absolute ideas may be known to remain supreme.[6] Plato's understanding was that the immediacy of the archetypes of the metaphysical world of divine reality, ordered the cosmos with its pervading meaning, bringing feelings of joy, love and peace to serve directly the interests of the soul. However, these states of being are in tension with the Reason of the natural philosophies defined by Aristotle. His definition of the role of philosophy is known to us from parts of his surviving work *De Philosophia*. The move to study the divine philosophy of formal and final causes and discover the intelligible essence of the universe behind all change arises after natural philosophy discovers the

---

[4]   Richard Tarnus, The Passion of the Western Mind, p. 19.

[5]   Sogyal Rinpoche, The Tibetian Book of Living and Dying, p. 187.

[6]   Richard Tarnus, The Passion of the Western Mind, p. 53.

material causes of things.[7] The legacy of these great minds created a polarity between the metaphysical archetypes and the rational sciences based on the five human senses. This created a polarity in the mind caught between life's inner and outer worlds, 'to be or not to be'.

For some additional ideas about this existential tension at work, we will consider an aphorism from the pre Socratic philosopher Heraclitus, c. 535-475 BC. He cites the example of a bow and its arrow: the polarity between the energy of life in the bow and the final purpose of the arrow, death. We know how a thirst for knowledge creates tension and drives individuals wanting to learn. The Greek philosophers that inherited the tension between the spiritual philosophy of Plato and the empirical approach of Aristotle were not satisfied until Plotinus c. AD 204-270 resolved this difficulty with his philosophy of Neoplatonism. During these five hundred years after Aristotle, the growth of spiritual intelligence faced many other challenges.

A student of Aristotle, Alexander the Great (356-323 BC) successfully conquered much of the Near East to spread the Hellenistic culture. Before his early death, he built the Egyptian coastal city of Alexandria as the centre for the refinement and dissemination of Greek knowledge. Eminent men of philosophy and literature gravitated from the surrounding lands to establish amongst its schools an excellence known as the Alexandrian School of Philosophy. It was in Alexandria that the teachings of Jesus c. AD 30 and the universality of the Christian religion were considered by many philosophers alongside the exclusivity of the Jewish faith. The Greco-Roman culture of the Roman Empire absorbed Christian philosophy and facilitated the teachings of the Church of Rome.

The Neoplatonism of Plotinus was based on three principles: the One, the Intellect, and the Soul. His distrust of the human

---

[7] Ibid, p. 66.

body and normal materiality stemmed from his belief that the physical dimension of life is only a shadow of 'higher and intelligible' realms experienced within the 'truer part of genuine Being'. These themes captured the minds of the important third century Christian thinkers including Clement of Alexandria, Origen and later Augustine of Hippo, 354-430. During the centuries when orthodox Christian doctrine was being established it is of interest to consider the degree of spiritual intelligence then expressed. The tension between the individual's inner and outer worlds was resolved through the Christ energy's expression through the ministry of Jesus thus rendering philosophical enquiry redundant. However, the Church's doctrine of Original Sin created and taught of a new source of tension to deny each individual's personal inner source of God's unconditional love and compassion.

From around the year 610 until his death, Muslims believe Muhammad's teachings came from the prophetic Messages given by Allah, which form the basis of the Koran. Muslims identify the Prophet Muhammad as the last of God's prophets since they believe the Christian Church strayed from God's path. The practice of Islam as revealed by Prophet Muhammad is simplicity itself; a believer worships God directly with the sacred chant - There is no god but God and Muhammad is His Messenger - without the intercession of any worldly authority.[8] Before his sudden death in AD 632 many tribes from Arabia came to support Prophet Muhammad's religion, the worship of Allah. For a time this put an end to all Arabian tribal wars and united them within the Kingdom of Islam - a common religious and political system. From Arabia Islamic armies quickly conquered the nearby lands of North Africa and the Middle East. Islam's monotheistic religion expressed tolerance towards Christian and Judaic believers.

[8]   Jonathan Bloom and Shelia Blair, *Islam Empire of Faith*, p.35.

During the next 300 years the Golden Age of Islam provided an unprecedented growth of scholarship with Arabic the international language of the Divine Truth. In Baghdad, at Bayt al-Hikmah, 'the House of Wisdom', scholars translated much of the surviving philosophical, scientific and medical works of the ancient world. In addition to enriching Islamic culture, its scholars also extended this knowledge. The destruction by the Mongols of much of what the Muslims had accomplished by the end of their Golden Age was a great loss.

Resistance by the Church to the spread of Islamic beliefs reached a peak with the work of the Frankish King Charlemagne (741 - 814) when his army established Western Christendom in Europe. The Christian corridor linking the British Isles and Italy stretched from the Atlantic to the Danube. This region, created and secured by the sword, was a Christian stronghold against paganism and the armies of Islamic believers. Intrinsic to Charlemagne's effort was the determination to create the conditions for the acceptance of a new European cultural life independent of the flourishing Byzantine culture of the Eastern World. To create this new Frankish court Charlemagne gathered many learned men around him that were to provide the scholarship and culture that created the Carolingian Renaissance of this period. Alciun, the leading ecclesiastical scholar from York was invited to set the syllabus and teach at the court school of Charlemagne. At this time the Church focused on traditional Christian learning and the monasteries welcomed many sacred Christian texts retrieved for copying and study. Alciun's writings provided a significant contemporary record of the 793 Viking attack and massacre at the Lindisfarne church of St Cuthbert.

Viking raids along Europe's north western coast began c. 700. During the next two centuries the lack of land alongside the fiords of their homeland motivated generations of young men to be more ambitious as these attacks penetrated further inland with the invaders creating permanent settlements. Despite the carnage caused by their arrival they were free of the cultural constraint imposed by personal land ownership. The Vikings introduced a notable cultural achievement recognised as a vital step towards Europe's modern democratic assemblies. The early popular Scandinavian meetings of local chiefs 'appointed judges, made laws and executive decisions.'[9] The historical record of ninth-century Sweden describes these local gatherings as the *ding*. Denmark and Norway had their *landings* and *logthings*, regular meetings to consider the interests of individuals. This change developed the idea of the rights a person had. These were beyond the cultural practices of ancient Greece and Rome and threatened the primitive culture of 'might is right.'

During the centuries that followed, where local native populations recaptured control of land from Viking invaders, a form of feudal culture developed based on the obligations that arose from the gift of land from the King to those who had served him well. In return for this benefit landowners had to provide protection, such as a mounted knight for their kingdom. These new landowners rented out their acreage to tenants who were then obligated to pay their landlords a portion of the produce harvested. These legal bonds greatly reduced the individual's freedom at all cultural levels. Alongside this allocation of productive pastures, the monasteries had access to marshlands and other areas difficult to farm where the hard work of monks created local centres for trade. This success built on the compassion expressed by monks towards the needy souls of their locality.

The education of monks was expanded by the monastic study of early Christian documents. These groups widened to include secular aspirants. In 1130 Paris, at the Augustinian

---

[9] *Europe A History*, Norman Davis, p.297.

Abbey of Saint-Victor, the monk Hugh of Saint-Victor expanded education to include the secular education that focused on the seven liberal arts: grammar, rhetoric and dialectic (known as the trivium) and arithmetic, music, geometry and astronomy (known as the quadrivium).

The growth of spiritual intelligence within the different cultural contexts of monotheism, the rights of the individual and those of the crown created many tensions within and between cultures. The challenges that arose to reduce these tensions created major agendas that continue to test the spiritual competence of all concerned. The first of these we will look at concerns those rights of the individual, which were part of the culture the Vikings introduced.

Within a few centuries, Scandinavian ideas influenced the Western European legal standing of the individual. The extent and importance of this issue is illustrated in the signing of Magna Charta in 1215. The impact of growing numbers of individuals with the freedom and aspirations to pursue their own interests extended beyond the ownership of land and the power of the crown. We can trace this progress from within Charlemagne's monasteries where the scholastic initiative of the monks was exercised by their spiritual guidance. This had started to include the study of the classical Greek spiritual philosophy so thoroughly assessed and debated in Roman Alexandria, c. AD 150 Egypt. Such matters spawned contemplation and debates during the centuries that followed and created an iconoclastic threat from within the Church itself. Until the Renaissance, Christianity had to function within the disciplines designed to restrict the spiritual pluralism necessary for spirit to motivate an individual's life.

Aspirations to use spiritual intelligence for the monastic study and contemplation of God reached its peak with the work of the mystic Thomas Aquinas (1225 - 1274) whose initiative included the study of nature as an aspect of God's Creation. By this time the Church had obtained many manuscripts from classical Greece. While dismissing pagan beliefs concerning animal and nature spirits he nevertheless embraced the study of the natural world by arguing that its wonders are further evidence of God's divine omnipotence. Each species, irrespective of its miraculous natural qualities, depends on the greater intelligence of Life for its own existence to be supported. Aquinas redeemed the fall from grace that the natural world and its believers had suffered in Christian eyes under the pen of Saint Augustine (354 – 430) who argued that human nature is inherently full of sin from birth.

Christian theologians had long pondered over the wisdom provided by the writings of the great Greek philosophers Plato and Aristotle despite their differing emphasis on human mystical and rational experiences. Aquinas showed how a Christian philosophy was able to interpret the work of each of these sages as complementary to each other, e.g. Aristotle's empirical measurements of particular aspects of life were deepened through an awareness that their common existence depended on the part each species played within the greater environment. The importance of particularised or individual aspects of life lie entirely on how their archetypal qualities interconnect with each other and create the dynamics of all life's relationships. The divine intelligence of Creation includes an all-embracing web of life's species. Knowledge of all these archetypes could only expand Christian wonder and faith in the Creator's work. Aquinas argued how the ideal of secular reason and religious faith enrich each other and serve the appreciation of life's deeper unity.

This challenge to Church doctrine was a harbinger of the many facets of the Renaissance about to explode the cultural conservatism established by the Church. The achievement of the Renaissance introduced greater cultural diversity and created the momentum powering the process that developed through the Reformation, the Scientific Revolution, the Enlightenment, and the Industrial Revolution of the eighteenth century. The environmental and social costs demanded by industry's hunger for labour and raw materials were recognised in the aspirations of the many individuals that voiced the ideals of the Enlightenment. The tension between the use of industrial power and ideals of the Enlightenment echoed the polarity between the individual's inner and outer worlds as expressed by the Romantics. Blake's 'dark satanic mills' are emblematic of the industrial exploitation of the poor and nature's resources. Christian ideas about compassion required practical attention. These included the 1833 ending of the British Empire's participation in the Slave Trade.

The results of cultural change, from the power of an enforced religious Church doctrine to power based on scientific reason, have polarised the extremes of human interests and had huge social and ecological repercussions. This is not a reflection of a decline in spiritual sensitivity but rather the expansion of the trained secular mind and its aspiration to seek inspiration about any subject that people aspire to understand. This spiritual opportunity has been used by science in its search and discovery of the knowledge needed by modern industries.

Of the many philosophical theories of knowledge advocated down the centuries, the scientific advances of recent centuries support the idea of the universal mind proposed by Bishop George Berkeley (1685-1753) as providing the source of all knowledge available to anyone able to access it.

The experience of being open to the mystical dimensions of life to solve personal mundane challenges uses the polarity between the inner and outer world identified in the way the works of Aristotle and Plato may complement each other; often at the cost of increasing tension between narrow and specialised interests, both personal and cultural.

From this overview of the growth and use of spiritual intelligence, we recognise that for new spiritual ideas to be accepted individuals were required to have the aspiration and initiative to succeed in their search for answers. These ambitions were usually expressed after some form of education enabled these people to establish positions within their communities from which they were able to express the unique originality for which each is recognised.

7

Patterns within the Growth of Intelligence

In recent Chapters we looked at how different aspects of intelligence emerged within different cultures during the last twelve millennia. From this overview of our emerging consciousness the historical evidence of the main themes from Egyptian, Greek and European cultural development follow a common pattern. In each case around six centuries after a new theme of excellence was experienced, it matured into wider cultural expressions. In this chapter, we will first look at these examples and then see what prehistoric evidence can be found.

To begin we will look at the time around 3200 BC when the Egyptian nation was being formed. The introduction of the Sun cult aided the unification of the Upper and Lower kingdoms of the Nile valley. Egypt's national cohesion depended on the continuity of the spiritual integrity of kingship maintained over later centuries to bring blessings to the Egyptian people. Each king was required to combine divine intelligence with his mortal responsibilities. The use of hieroglyphs introduced mental intelligence that identified the spiritual symbolism of the name Horus; the falcon head on the tombs of their kings. The success of Egyptian royalty is recorded in the evidence of their famous pyramid culture and other achievements from the period of the Egyptian Old Kingdom, which began c. 2,600 BC.

The second example concerns the Greek influences behind the expansion of education. The cultural place of the hero was immortalised by Homer's c. 900 BC classics describing the victorious Greek heroes at the battle for Troy. Their exploits focused attention on the archetypal qualities needed by the individual to receive their culture's accolades. Some 600 years after Homer's *Iliad* and *Odyssey* had established the archetypal foundations of Greek culture, Aristotle, 384 - 322 BC introduced more realism. He is recognised for the sharp differences between his philosophical stance and the spiritual archetypes favoured by his master, Plato. His preference for an empirical approach to understand life laid the foundations for our modern universities.

The third example of this 600 year phase of development is seen in the years between c. 1150 and c.1750. It concerns the growing aspirations that became possible after the cultural introduction of education. It began earlier when the Church started to teach their monks Latin so that they could study the holy manuscripts Charlemagne had collected for his monasteries. This initiative culminated with the acclaimed scholarship of Thomas Aquinas. This introduced an understanding that enabled the Church to embrace rather than feel threatened by Greek philosophy. The culture of the Middle Ages expanded when the Church accepted secular students into their schools. During the later Renaissance, the spiritual contributions from the protestant monk Luther and the science of Copernicus helped the wider use of knowledge nurture powers that rivalled those of the Church. The discoveries of Kepler, Galileo and Newton contributed to the later Scientific Revolution. By c. 1750, the Industrial Revolution had begun with water mills powering the early factories. Later steam engines powered industries across England. George Stephenson and his son Robert's steam locomotive, the Rocket, established their practicality. This led to the growth of railways, vital for the transport of industrial raw materials and finished goods.

During the prehistoric millennia before 3,500 BC, the record of the emergence of new expressions of consciousness is meagre. One sphere of activity where a new type of intelligence was expressed is illustrated by intuitive skills developed for horse riding. The emotional intelligence for the domestication of the horse is estimated to have begun after c. 5500 BC.

Over some two millennia, traders had much experience with the movement of merchandise over long distances to bring goods to the market place. The earlier challenge to move food between local communities had become part of normal life after the farming revolution created the start of the Neolithic Period. Around this time, the use of boats had become part of northern European culture.

Table 7 below shows that the times when new emotional mental and spiritual dimensions of intelligence were being introduced were followed by the aspiration to use this intelligence further. These three processes spanned the years from around 3000 BC to AD 1800 and each of these strands of progress included the 600 year phase. This period passed between each new dimension of intelligence being introduced and the time when each new skill became widely practised. After being established in their native culture, they were then widely disseminated.

Where historical facts are available, this pattern suggests that key individuals initially cultivated new ideas, which were later expanded to serve the common good. This process probably relied on the contributions of wise men or 'sages' expressing and introducing new qualities of intelligence within an elite group. During a period of some 600 years, these ideas matured and permeated their respective cultures to express new peaks of creative excellence.

Through the millennia spanning the emergence of physical, emotional, mental and spiritual aspects of intelligence, the arrival of each new quality provided the potential to extend and refine the ways earlier aspects of intelligence had been used. These later phases of the greater expression of intelligence are shown below.

Table 7

Contexts for the Four Phases of Growing Intelligence

Connections	Expressions of New Intelligence		Later Cultural Expansion	
The Survival of populations	The Neolithic Revolution. Crop cultivation began. The invention of boats.	from c. 10,000 BC	Regional spread of farming from Jericho. Limited evidence of boats in use.	from c. 9,000 BC
1. Physical movements	The sharing of surplus grain. The movement of traded goods.	from c. 8,000 BC	Expansion of villages as towns became centres for trading.	from c. 7000 BC.
2. Emotional intuition	The individual's personal skills in cultural use. The taming and riding of the horse.	from c. 6,000 BC	The regional power of Mesopotamian City States. The early use of wagons.	from c. 5000 BC
3. Mental education	Hieroglyphs of Egypt and Mesopotamia. Religion unifies Egypt's Upper and Lower Kingdoms.	from c. 3,500 BC	Birth of Egypt and the start of Egyptian Old Kingdom with its famous pyramid culture.	from c. 2660 BC
4. Spiritual inspiration	Greek heroes of Troy. Homer's poems tell of archetypal heroes.	from 1,000 BC	The heroes of the Golden Age of Athens. Socrates, Plato and Aristotle.	from 400 BC
Aspiration to use spiritual intelligence	Church schools start to accept secular students. Aquinas argues Christian theology and Greek metaphysics both serve to appreciate life's glory.	from AD 1,000	The power of the Church had declined from its loss of spiritual Integrity. Man's 'reverence' of his power to control nature became a 'sacred' culture.	from AD 1500 To 1800

An early example of a new skill enhancing the development of an earlier advance arises when we consider the physical movement of goods. We know how trading benefited from the domestication of the horse. New mental skills enabled different cultures to communicate better with each other for their mutual benefit. Spiritual intelligence developed in ways that used writing to move ideas about. A feature of all these advances was the gradual increase in the numbers of people included in the new ways intelligence emerged.

They show how after the Ice Age ended and a new physical survival culture had been established new dimensions of intelligence began to emerge. Table 7 shows the four 2000 year periods during which movement, when addressing physical, emotional, mental and spiritual agendas, contributed to the growth of intelligence. The final period identifies the context of a new spiritual quality of aspiration beginning to create our modern culture.

An observation concerning the iconic buildings that have come to symbolise three empires reveals a strange correlation with periods of some 2000 years. Close to the modern Egyptian city of Cairo, the Great Pyramid of Giza was apparently built c. 2560 BC as a monument to Pharaoh Khufu for successfully maintaining the unification, security and prosperity of the tribes of the Nile valley.

On the Acropolis of Athens 2122 years later, the famous Parthenon was completed, c. 432 BC. This building was the temple of the goddess Athena with the famous words 'Know Thyself' carved over the entrance to provide an example of spiritual intelligence. All human skills benefit from the wisdom of this goddess, her attributes associated with the night vision of the Little Owl often illustrated in the company of this goddess.

Over 2146 years after the Parthenon was completed, in the city of London the new St Paul's Cathedral was consecrated in 1708. At this time the city of London became the centre of aspirations for global trade.

The chapters exploring our emerging intelligence have identified patterns that contributed to establishing the roots and enhancing the expansion of past civilizations. This observation prompts the following question. Together, are the near 2000 year time spans in the table above one part of a much longer process that is gradually manifesting new qualities of human consciousness? If this is the case, other human qualities are likely to be subject to their own sequence of development through physical, emotional, mental and spiritual stages. To assess this possibility we will look for natural phenomena to see if we can find a model to evaluate this idea.

8

Themes from Past Millennia

The idea that intelligence has taken some 10,000 years to progress to its current level takes a while to appreciate. We are more familiar with thinking in terms of recent centuries for technologies to develop. What other evidence is there that natural change needs millennia for results to be recognised? Concerning geological changes, our current Holocene period, which began 10,000 BC, followed the Pleistocene epoch. This is the latest of the many geological periods within the last 600 million years. When studying the Earth's climate each of the successive Ice Ages define another order of time. Recent Ice Ages lasted some 100,000 years and were followed by a short warm inter-glacial period before the next Ice Age arrived.

The last Ice Age reached its coldest at the last glacial maximum (LGM), some 20,000 years ago. The rate at which the ice melted over the following 12,000 years was irregular, the warming environment allowed the human population to grow. Numerous regional populations adapted to these changes according to their particular challenges and opportunities. The new activities introduced since the LGM have been ordered according to human tool making skills. These identify the Stone Age, its later Mesolithic and Neolithic Stone Ages, followed by the Copper, Bronze and Iron Ages.

For some other ideas about past human ages, we have the work of the c. 700 BC Greek writer Hesiod to consider. He wrote of their identifying characteristics by using two myths. The first was the story of Pandora, who opened a jar of evils into the world. The second concerned the decline of humanity from a distant Golden Age through the Silver, Bronze and Iron Ages down to his own Heroic Age. He traced this 'decline' of humanity in his book of wisdom titled *Works and Days*. He identifies the twin virtues of hard work and honesty as the means of self-improvement, when energies are focused on achieving worthwhile and lasting goals. During the mythic original Golden Age, 'the first men enjoyed complete happiness. They lived like gods, free from worry and fatigue; old age did not afflict them; they rejoiced in continual festivity. All blessings of the world were theirs: the fruitful earth gave forth its treasures unbidden. They died as overcome by sweet slumber to become benevolent genie, protectors and tutelary guardians of the living.' 'In the Silver Age (a period of matriarchy) there lived a race of feeble and inept men who obeyed their mothers all their lives'. 'The men of the Bronze Age were as robust as Ash trees and delighted only in oaths and warlike exploits. Their pitiless hearts were as hard as steel; their might was untameable, their arms invincible'. Hesiod associates the final Heroic Age with the valiant warriors who fought before Thebes and under the walls of Troy.[1]

The mythology of Hesiod concerning the Ages of Man, which echoes the year's four seasons resonates with myths from several ancient cultures. Writing about the Cycle of the Ages in his book *Lost Star of Myth and Time* Walter Cruttenden reviews Hindu, Vedic, Egyptian and Mesoamerican belief systems with their own spiritual understandings of past and future Golden Ages.

Could the emergence of the different facets of intelligence earlier identified as a process taking place during past millennia somehow relate to these past understandings of the ages? If nature is our greatest teacher, where in the natural world can we find some clues about past ages to help us rationally assess this possibility? We will start at a basic level by considering how life creates growth in nature. Is there a definitive process on our planet that governs how life manifests in our physical

[1]   Guirand, F. *New Larousse Encyclopaedia of Mythology*, p.93.

realm? At a fundamental level, the seasonal variation in the energy each region on Earth receives from the Sun drives our global weather systems. The Sun's annual cycle has successfully provided the source of energy necessary to establish and renew the countless elements in the food chains needed for life to continue and evolve over millions of years. Although the annual growth of food for countless members of different species is not in doubt, how many will survive each winter is uncertain.

Celestial Cycles

How is it possible to relate the annual cycle of the Earth's orbit around the Sun to the much longer periods with which our study is concerned? The use of celestial cycles to bring some sense of order into the management of time has been practiced for many millennia. When it started is unknown. The annual cycle of star patterns in the night skies could have been used by hunter/gatherers to time the migration of reindeer, a staple food for northern tundra Ice Age communities. A carved engraved bone plaque dated c. 33,000 BP (before present radiocarbon date), found in the Blanchard rock shelter in France has a pattern of dots that suggest an interest in lunar astronomy.[2] The helical rising of Sirius, the brightest star was used 5000 years ago to forecast the Nile's annual flood. The Moon's phases and monthly cycles still provide an objective method for people to organise their lives. The extent of this once popular tradition can be gleaned from the book *The Art of Timing* by Paungger and Poppe. This work considers the application of lunar cycles in daily life used today.

Astronomical measurements of Space and Time.

For calendars of longer periods there is evidence from Babylon of two astronomical tablets, (mul.apin) dated 687 BC listing 12 thirty day months, 18 bright stars and the belt of

---

[2] Paul Mellars, *Prehistoric Europe – Upper Palaeolithic Art*, p.71.

constellations.[3] From the same region during the fifth century BC precise astronomical information was recorded on Cuneiform script tablets.[4] These record the degree and minute positions of solar eclipses in the zodiac constellations. Apparently, it has been possible to distinguish the differences between observed and calculated positions in space.

The studies of these fifth century BC astronomers focused on the movements of the Sun and planets within the background belt of unequal constellations. This belt is about 16 degrees wide and when divided into twelve equal 30 degree arcs becomes the sidereal or Babylonian zodiac. The reference axis for measuring the positions of the observed planets and stars is provided by the axis between the stars Aldebaran and Antares. This selection was based on their positions opposite each other neatly dividing the heavens into two very nearly equal halves.

Babylonian mathematical astronomy with this zodiac was different from the contemporary practice of Greek astronomy. Using the positions of stellar features for measuring time was a practice refined by the astronomers of Athens. A description from the fifth century BC illustrates their identification of the beginning of winter with the evening rising of the star Arcturus. For measuring the length of the year *The Seasonal or Natural Calendar* advocated by Euctemon was popular with its 7 thirty day and 5 thirty-one day months.[5] However, new or full Moons could not accurately forecast the Sun's annual order of time, which dictates the traditional seasonal festivals.

A problem was discovered when Greek astronomers were exploring for solutions to the difficulties caused from using their traditional lunar calendar. The need to replace them arose because of the political and economic problems caused when

---

[3] Robert Powell, History *of the Zodiac*, p.109.
[4] Ibid, p.39 & 40.
[5] Pritchett and van der Waerden,
   Thucydid Time – Recognising and Euctemon's Seasonal Calendar.

the dates of the Roman Empire's religious and secular festivals failed to maintain an alignment with the climax of each season. The divisive solution to this problem gave rise to the 'movable feasts'. The power of the authorities to change the traditional calendars to maintain this alignment risked being abused. For scores of millennia the lunar calendar provided a reference for cultures to use in relation to many of life's experiential agendas. This simple way to count the weeks and months had a consistent error when used to understand the seasons of the year. Creating a solar calendar also had difficulties.

Annual cycles of the Sun and the Great Year

To appreciate one problem, we need to look at the finer details of timing when measuring the precise duration of a year. In fact, there are two valid ways to measure this period. Firstly, the solar year measures the time for the Earth to orbit around the Sun. Traditionally, this year is measured at the spring equinox. The time in the year when the annual northwards motion of the Sun's position is directly above the equator. The second method measures the sidereal year, the time for the Sun to return to a reference point as it sweeps around the heavens against the background patterns of stars. This sidereal year is about twenty minutes longer than the length of the year taken for the Sun to return to the same position relative to the Earth.[6] This means that at the end of a solar year, the year measured with reference to the stars (the sidereal year) has not yet finished. Because of this twenty-minute difference, each year the sidereal reference point in the heaven moves backwards through the stars creating the movement known as the precession of the equinox. This motion of the spring equinox through the constellations is in the reverse order to that observed when the Sun passes through the constellations during the months of the solar year. This rotation is caused by

---

[6]   David Ewing Duncan, *The Calendar*, p.324.

gravitational influences on the Earth from the Sun, Moon and the planets of our solar system. After an estimated 25,800 years, the position of the spring equinox passes through all the constellations of the zodiac. During this period, the vernal reference point moves backwards through one complete circle of the heavens.[7] When this period of 25,800 years is divided by the cycle's 360 degrees we find that each degree of backwards movement takes 71.667 years to complete, and provides us with an estimate for the rate of precession. Small variations in this speed are caused by changes in the gravitational influences that create this movement.[8]

Understanding the implications from discovering the two ways of measuring the length of the year enabled an accurate solar calendar to be devised. The solar year measures the time taken for the Sun's cycle around the ecliptic. The sidereal year depends on the cycle of the Sun around the background pattern of stars. The Greek astronomer Hipparchus (148-127 BC) is attributed to have discovered a difference between these two periods, the scientific evidence of precession.[9]   The details, which Hipparchus is likely to have used to discover this fact is included in the book, *History of the Zodiac* by Robert Powell.

It is probable that Hipparchus recognised that any calendar based on the sidereal zodiac, although more reliable than a lunar calendar, would lose its seasonal accuracy. To solve this problem, he defined the Greek or tropical zodiac for the solar year which came into use after the second century BC.

A modern way to observe precession is to display this motion with astronomical software. When viewed on the computer screen the software that simulates this precessional movement can speed up time for the observer to appreciate the

---

[7]   Jacqueline Mitton, *Penguin Dictionary of Astronomy*.
[8]   In *The Sidereal Zodiac,* p. 44, Robert Powell notes that the current rate of
      precession is 1 degree in 71.63 years.
[9]   Robert Powell, History of the Zodiac, p.37.

movements century by century. Using this means of studying the heavens at the spring equinox in annual increments, the movement of a constellation just above the eastern horizon, before sunrise, reveals the movement of the zodiac. Over the millennia, each of these constellations moves closer to the eastern horizon before disappearing. During the last 2000 years, the visible end of the constellation of Aries the Ram and most of the two fish of Pisces have gradually slipped backwards and out of sight on the eastern horizon at the spring equinox. During the next 300 years, the remaining few degrees of Pisces will disappear. At this time, the end of the constellation of Pisces denotes the beginning of the next 2150-year astronomical Age of Aquarius.

The mode of rotation of the Earth that causes precession is illustrated by a child's spinning top as it gyrates just before it falls over and stops rotating. See Fig 8.1.

Fig 8.1

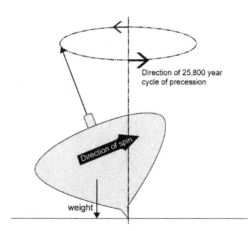

When the top is spinning fast its axis of rotation is vertical. When it slows down, before it topples over and stops, its axis of rotation begins to gyrate sweeping out a cone shape in the air above the top.

This gyration illustrates how the Earth's axis between the North and South Poles gyrates at an angle of 23 degrees and 27 minutes sweeping out a cone shaped space during each cycle of precessional rotation. Each Pole points towards a place in the heavens called the northern or southern celestial pole and each traces a circle amidst the stars every 25,800 years. See Fig 8.2 below.

Fig 8.2
Circular Path of the Celestial North Pole in the Northern Sky

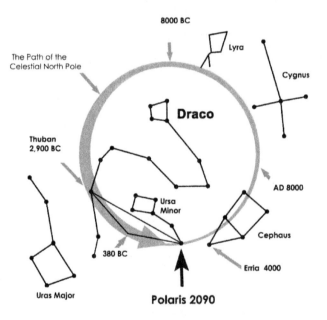

Close to the end of the 21st Century, the point in the northern heavens, the celestial North Pole, will be at its closest point to Polaris, our current northern Pole Star. Around 2,900 BC the celestial North Pole was approaching Draco the alpha star of The Dragon constellation.

When the 25,800 year the cycle of precession is divided into twelve 2150 year astronomical ages these represent the months of the Great Year.[10] These divisions present a scale of time used to identify the astronomical ages.

This scientific understanding had repercussions for the Mediterranean religions dominated by cultures believing the wishes of the traditional gods and goddesses could be understood through the art of astrology. With the new knowledge of precession, a god had been identified that was so great he controlled the movement of the very heavens wherein all the traditional deities resided. A god with such omnipresence introduced understandings that all change flowed from this higher source.

In his work *The Origins of the Mithraic Mysteries: Cosmology and the Salvation of the Ancient World*, David Ulansey writes that appreciating the astronomical discovery of the precession of the equinox provides the key to understanding the religion of Mithraism better.[11] At one time, it rivalled Christianity as the preferred religion of the Roman Empire. The secrets of this cult (c. 70 BC–c. AD 400) have been difficult to penetrate since the beliefs associated with this religion were never written down. The discoveries that have been made are the result of extensive detective work.

Evidence of its popularity outside the society of the cultural elite comes from the hundreds of Mithraic temples used by soldiers across the Roman Empire. These have provided a valuable source of material from which scholars have interpreted some of the cult's secrets. Common to all temples was the central image of Mithras slaying a bull, the tauroctomy. This 'altar' was always in the symbolic company of a dog, a snake, a raven and a scorpion. Recent work on understanding

---

[10]   Western history records how ancient astronomers identified other long cycles as a Great Year. See *The Great Year,* Appendix 4 by Nicholas Campion.
[11]   David Ulansey, *The Origins of the Mithraic Mysteries: Cosmology and the Salvation of the Ancient World,* p 82.

the significance of these creatures suggests that they represent an astronomical map of several constellations. Those related were Taurus, Cannis Major, Hydra, Corpus and concluding with Scorpio (the sign opposite Taurus).[12] How this symbolism has been interpreted relies on an understanding of precession. As the constellation of Aries denoted the position of the spring equinox during the era in question, the constellation of Taurus related to the position of the spring equinox during the previous age. Within this context, the worship of the god Mithras slaying the bull can be interpreted as devotional practices that honoured the divine power that controlled the changing of the ages.

Before the ideas of precession were linked to the Roman Mithraic beliefs, this cult was thought to have developed from the myths of the great Persian god Mithras. In pre-Zoroastrian times, he was related to the god of light, Ahura Mazda. This background knowledge supports a theory for the great secrecy surrounding Roman Mithraism. This was that the beliefs of this religion were contrary to the prevailing scientific wisdom and threatened to create a cultural division between science and religion. If the god of light is the same as the god of time, another interpretation of the tauroctomy arises. Mithras becomes an image for the eternal Man. He is sacrificing his powerful animalistic drives (related to the Bull of Taurus) to experience enlightenment during the 'age to come'.

During its formulation in the early centuries of the first millennium, the Christian religion had to compete with Mithraism and other religions of the time. The central idea of an omnipotent Deity that presided over all changes and was able to intervene in worldly affairs was central to its success. It also provided a much simpler alternative to following all the practices demanded by a complex pantheon common to traditional pagan customs.

---

[12]   Different methods have been used to define the boundaries between the constellations during different periods.

Other cultures also studied the sky and invented zodiacs to suit their purposes. The Upper Egyptian town of Dendera is famous for its temple of Hathor not least for its zodiac ceiling dated to Greco-Roman times, which is currently in the Louvre. This later zodiac shows the constellations the Egyptians used spread across the whole vault of the heavens. This representation of the patterns seen in the stars illustrates how Egyptian astronomy and astrology were different from the Babylonian and Greek methods studied by philosophers in the schools of Alexandria, the centre of Mediterranean culture after c. 300 BC.

In Chinese mythology, twelve animals are used to create their zodiac. In India, the zodiac astrologer's use has its own history. It seems that the intellectual differences between these zodiacs do not reflect their usefulness. Their common feature of providing a holistic overview of life suggests their success is related to the quality of the motivation behind their uses.

The Astronomical Ages

We have scientific evidence for twelve near 2000 year astronomical ages during which numerous new aspects of human consciousness emerged. It is worth assessing if this cycle's symbolism correlates with the psychological changes identified to have taken place as intelligence developed in the near 2000 year periods identified. To consider this symbolism, we will look at zodiac mythology. In Hellenistic Alexandria, Greek astral beliefs and astronomy provided the basis for a new development in astrology.[13] The principle of correspondence was believed to relate zodiac symbolism to the behaviour of pagan gods and goddesses. Observations of many natural phenomena associated with the year's months are another level of correspondence described by the signs of the zodiac: themes to consider in the context of the months of the Great Year.

---

[13]   Robert Powell, *History of the Zodiac*, p.112.

When considering the validity of this approach it would be useful to know how much the Stone Age mind was influenced by the qualities of the year's changing seasons together with all the opportunities and challenges that these changes presented. It is hard to imagine that they did not significantly influence the development of all languages.

With regard the use of the symbolism of the constellations the Sun passes through each year, any chance of using this imagery in a precessional context and correlating it with the archaeological and historical records depends on accurately aligning the precessional cycle's symbolism with the dates of vital events from this study. This introduces a technical matter about the date when the cusp between the sidereal signs of Aries and Pisces, in the sidereal zodiac's belt of stars, was aligned with the spring equinox (or vernal point) of the Sun's annual cycle. The work of Cyril Fagan recognises that in the sidereal zodiac of fixed stars, they are not really fixed. By allowing for their slow motion (at a rate of 3.63 seconds per century), Fagan shows 'March 31, AD 221 as the date upon which the vernal point lay at the zero point of our formally-defined sidereal zodiac'.[14] This date allowed astronomers to identify the dates and symbolism for each 2150 year astronomical age. We will use these ages to consider the millennia since the ending of the last Ice Age. With this in mind, the commencing date of the astronomical ages is shown in Table 8 below.

Table 8.

The Astronomical Ages.

Cancer	Gemini	Taurus	Aries	Pisces	Aquarius
8,379 BC	6,229 BC	4,079 BC	929 BC	AD 221	2371

---

[14] Powell and Treadgold, *The Sidereal Zodiac*, p. 45

The Greek alignment of the cusp between the sidereal signs of Aries and Pisces with the vernal point (spring equinox) was a scientific measurement made to identify the start of the tropical zodiac. This basic reference point was chosen without any wider philosophical considerations. For insights on this matter, we refer to the work of Claudius Ptolemy, the famous astronomer from second century Alexandria. He published a star catalogue to formalise the Greek or tropical zodiac. This astronomical zodiac was used until the twentieth century. His books entitled *Almagest* and *Tetrabiblos* included much of the work of his predecessors. He writes 'The vernal point was adopted in Greek astronomy as the beginning of the ecliptic, apparently because the vernal equinox was considered by Greek astronomers as the start of the year'.[15]   This choice seems rather arbitrary from a philosophical point of view, since it was apparent that the equinoxes of spring and autumn, and the solstices of summer and winter, each have qualities that suggest they qualify for this distinction.[16] There is no hint that this alignment could relate to human development in any way.

This overview introduces the scientific roots and cultural significance of three different zodiacs, each of which is associated with the same symbolism. These are the twelve constellations of stars, the twelve thirty degree signs of the sidereal zodiac and the twelve signs of the Greek or tropical zodiac. When precession moves the position of the spring equinox backwards through the sidereal zodiac we identify the astronomical ages. Our detective work now is to see if the descriptions found of our emerging intelligence correlate with the zodiac symbolism used to define the recent astronomical ages. It is necessary for the symbolism to describe different qualities being expressed for each of the physical, emotional, mental and spiritual dimensions of experience.

---

[15]   Ibid, p.9.
[16]   Ptolemy, *Tetrabiblos I,* pages 59-61.

Part Two

Building and Testing

a Model for

Our Emerging Consciousness

9

The Expression of Natural Symbolism

From the detective work done in four earlier chapters we can appreciate how physical, emotional, mental and spiritual dimensions of intelligence began to develop after the last Ice Age. With the passing of these astronomical ages, we will now test a theory. Does the symbolism projected by ancient astronomer-priests onto these constellations correlate with the developments identified in these four chapters? This evidence provides the keywords aligned with the recent 2150 year astronomical ages shown in Table 9.1.

Table 9.1.

Relating the Astronomical Ages to Our Emerging Intelligence

Zodiac signs of recent conventional astronomical ages with the starting date for each age.	Intelligence Emerging when connections were made using different Dimensions of Experience		
	Dimension and *keyword* for quality of intelligence developed.	Physical dimension of experience and a *keyword* for the quality expressed.	The expression of new qualities of intelligence and the start of the age when it was first expressed.
Gemini Twins 6229 BC.	physical *connections*	physical *movements*	Trading of food and tools, c. 7500 BC.
Taurus the Bull 4079 BC.	emotional *intuition*	physical *resources*	Early use of cattle and horses, c. 5250 BC.
Aries the Ram 1929 BC.	mental *education*	physical *actions*	Cuneiform writing and hieroglyphs, c. 3300 BC.
Pisces the Fishes AD 221.	spiritual *inspiration*	physical *acceptance*	Spread of monotheistic religions, c. 1200 BC.

These ages follow the order defined by the movement of the precession of the equinox backwards through the zodiac. When each type of intelligence was explored, particular words are used several times since each captures an essential quality of symbolism being expressed. These correlations provide us with the *keywords* shown in Tables 9.1 and 9.2. Each describes a quality in approximate alignment with either the physical, emotional, mental or the spiritual dimension shown.

Table 9.2.

The relationships between the four stages of the growth of **intelligence** (in bold) within the Astronomical ages.

Signs of the Zodiac	The 2150 year Astronomical Ages			
	Gemini starting 6229 BC	Taurus starting 4079 BC	Aries starting 1929 BC	Pisces starting AD 221
Gemini connections	*physical movements*	*emotional intuition*	*mental education*	*spiritual inspiration*
Taurus fertility		*physical resources*	*emotional values*	*mental answers*
Aries strength			*physical action*	*emotional courage*
Pisces non-separation				*physical acceptance*

This table presents the four dimensions of intelligence horizontally; each roughly aligned with the symbolism of one the constellations, Gemini, Taurus, Aries or Pisces. By staggering the start of each constellation's contribution, we see how each may potentially relate to a dimension of intelligence known to have developed. We will now explore if the symbolism of these constellations correctly describes the qualities and dimensions of consciousness identified from history and archaeology. We recognise from this overview that the correspondence sought between the symbolism and when new changes took place, is some 1000 years out of alignment.

For each of the four expressions of intelligence already identified we will consider the symbolism of the four aligned constellations. To assess the accuracy of each of their correlations we will look at their symbolism in three ways.

1st context - within the Earth/Sun annual cycle.

2nd context - each constellation's star lore and its symbolism in a mythological context.

3rd context - to see if there is evidence describing the emergence of new aspects of intelligence.

When we look at this evidence the essential qualities of the symbolism are identified as keywords and shown in italics. This sequence is in the reverse order to that which describes the annual cycle of the seasons.

1. Astronomical Age of Gemini

The symbol of the Twins for the constellation of Gemini provides an image suggesting the potential for rich associations with the human condition.

The objective is to see if the emergence of new aspects of intelligence made during the Age of Gemini, 6229 to 4079 BC are described by this constellation's symbolism.

1.1. The symbolism of Gemini in the Earth/Sun annual cycle.

The zodiac constellation of Gemini provides the symbolism that spans the springtime month, from late May to late June in the northern hemisphere. What are the signs in the environment at this time of the year? How are these related to this constellation's symbolism? With the increasing warmth of the Sun, life is speeding up; many physical *movements* illustrate natural connections accelerating. We recognise the different *connections* various creatures make to find vital food supplies.

Why is this normal activity of unique significance during late spring? Is the survival of many species held in some delicate balance? At this time of the year, a gossamer thread holds the life of many new-born creatures. A familiar example shows how instinct is at work as young birds, after hatching, wait in their nests offering their open mouths to be fed by their parents. After the eggs have hatched sibling rivalry often results in the death of the weakest birds even without a cuckoo in the nest. How many worms and insects are sought out and brought to the waiting mouths in countless nests? Different categories of newborn creatures receive their nutrition in their own way; young mammals instinctively *move* about until they *connect* with their mother's teat to feed. Species from the botanical kingdom ensure their survival by providing the nectar needed by the insects that pollinate them. In this instance, physical intelligence is expressed as insects seek out life sustaining *connections*. The pollination of flowers during the spring starts a crucial stage of life's regeneration in one vital food chain. Many species migrate thousands of miles to rich feeding grounds to maximise the supply of food for the benefit of the new born. Instincts include the natural intelligence expressed as parents *move* about to make the *intuitive connections* their offspring depend on them for.

The particular behaviour of each species to ensure the survival of the next generation show us the wide range of strategies that are adopted. Different types of nurturing, like predators teaching their young survival skills are understandable, however the use of human physical, emotional, mental and spiritual dimensions of experience to describe what is happening in the non-human world is anthropomorphic.

1.2. The star lore of the constellation of Gemini and its mythological symbolism.

The star lore of this constellation has arisen from the names given to its two great white stars, Castor and Pollux, the heroic twins of Greek mythology. The brightness of Castor includes the light from a number of star systems including a spectroscopic binary and an eclipsing binary with at least six components altogether.[1] Pollux is a single orange coloured giant K star. These astronomical details reveal two contrasting stellar phenomenon coming together as the source of the imagery for this constellation's symbolism.

Greek mythology identifies the experiences of the twin brothers Castor and Pollux as heroes, providing many stories describing a wide range of experiences and services to humanity. The myths of their birth vary, one myth tells of Castor and Pollux being the sons of the king and queen of Sparta, Tyndareus and Leda. The 18th Homeric Hymn identifies the twins as the sons of Leda and Zeus and thus both qualify for the title Dioscuri, the immortal sons of Zeus – 'children who are deliverers of men on earth and of swift-going ships when stormy gales rage over the ruthless sea.' Were these waves also the white horses tamed by the Dioscuri, and which carried these heroes with the Roman cavalry for the 495 BC victory in the Battle of Lake Regillus? Greek and Roman temples in Athens and Rome testify to the twins' status as gods. Their worship and the invocations they *intuitively* attracted sought greater physical strength, speed and dexterity, which they famously gave to sailors, warriors, equestrians, athletes and others in need of exceptional skills. The myth where Castor is human and Pollux immortal arises after Castor is killed rather than be parted from his brother; their father Zeus allows them to spend alternate days in the Underworld and with him on Mount Olympus.[2]   The differences between their origins and the variations in their other myths may be appreciated as enriching rather than confusing their symbolism.

To obtain an overview of some of the many facets introduced by Gemini's twin heroes Castor and Pollux we will

[1]   Jacqueline Mitton, *Penguin Dictionary of Astronomy*, p. 62.
[2]   Kerenyi, *The Gods of the Greeks*, p. 108.

explore some of the perspectives about heroes considered by this symbolism in the writings of Joseph L. Henderson. His contribution to Part 2 of Carl Jung's *Man and His Symbols,* explores *Ancient Myths and Modern Man.* The insights concerning twins are one part of the descriptions Henderson provides about *Heroes and hero makers.* He presents evidence of different aspects of a maturing process from the work of Dr Paul Radin, *Hero Cycles of the Winnebago*, published in 1948.

This work identifies and describes four distinct hero cycles, the *Trickster* cycle, the *Hare* cycle, the *Red Horn* cycle and the *Twin* cycle. The *Trickster's* priority is to focus on satisfying basic personal needs. During the *Hare* cycle, giving an individual their Medicine Rite, he became their culture's hero/saviour.' *Red Horn* figures were super-heroes, invincible in battle; they proved themselves through superhuman powers – human weakness being compensated for by powerful tutelary gods. Lastly, the *Twin* cycle, 'Though the Twins are said to be the sons of the Sun, they are essentially human and together constitute a single person'.[3]

Henderson writes that the ego consciousness may mature and thus become more humble and be aware of its weaknesses as well as its strengths. The individual's initiations into different stages of their life may eventually result in them becoming a mature hero or heroine – to have completed the stages that enable them to express the creativity of their unique gifts.

This personal duality inherent in the symbolism of this constellation may be expressed by the masculine and feminine qualities within the individual's psyche. Cultural heroes or heroines could be the individuals whose wholeness is expressed with these polarities in balance. The individual's capacity to experience making *intuitive connections* within and between the physical, emotional, mental and spiritual dimensions of their experiences can use the symbolism of the twins of Gemini to help us understand a person's journey to

[3]   Joseph L. Henderson, *Man and His Symbols,* edited by Carl Jung, p. 106.

discover that their ego is not their Self. The versatility and ingenuity of the ego to justify its survival, challenges the observer within to be aware of the ego's limitless capacity to justify itself. The ego's imagination and speculations about life's looming possibilities focus on weaknesses and insecurities that expose personal vulnerabilities.

These different aspects of the psyche are symbolised by the very different astronomical nature of the two stars Castor and Pollux. Castor comprises of many complex astronomical systems whilst Pollux is a single star.[4]

The twins of the constellation of Gemini symbolise duality as depicted by its glyph, II. Two columns or pillars either side of an entrance or gateway denote the change from one place to another. *Movement* is necessary through a portal to make a connection between both of its sides. An exchange of energy may result at a recognised meeting place or a boundary where an agenda may be addressed. Two bronze columns were placed in front of Solomon's Temple. The one named Jachin 'he (God) establishes' was placed to the south; the one called Boaz 'by his (God's) strength' was placed to the north.[5]

The characteristics of a duality are described by the qualities and powers created when resonating energies are in proximity with each other (as with the polarity between the poles of a magnet). Movements between life's major polarities, love and fear, light and darkness, life and death, the ego and its projected shadow, consciousness and the unconscious and between genders present dynamics where the connections symbolised by the Twins may aid our deeper understanding of observed behaviour.

The Gemini duality includes the idea of the unusual combinations of ideas that can create both humour and trickery. To explore the extent of human deceit and betrayal

---

[4]   Jacqueline Mitton, *Penguin Dictionary of Astronomy*, p. 296.
[5]   *The Old Testament*. Kings, Chapter 7.

further we will look at the Scandinavian legends of Loki that feature in Teutonic mythology. The *New Larousse Encyclopaedia of Mythology* describes many stories where the name of Loki 'appears as often, if not more often than Odin and Thor'. 'Among the gods he was a sort of *enfant terrible*. He shared their lives and on many occasions zealously served them, and yet he almost never ceased working to undermine their power'. Eventually, Loki was murdered after he killed the god Heimdall. 'He was the god of light. His name probably signifies 'he who casts bright rays'.' 'No one in the world had an eye more piercing or an ear more acute than Heimdall, and yet he allowed his sword to be stolen by Loki -'.[6]

1.3. We will now consider the historical evidence for the emergence of new expressions of intelligence between c. 7,000 BC and c. AD 1000. These are described in chapters 3, 4, 5, and 6. In these chapters, illustrations are given of the 'qualities' of different types of new connections that changed the cultures identified.

The millennia spanning this period are about four times as long as the astronomical Age of Gemini, (the period between 6229 and 4079 BC). During the 2000 year period from c. 7000 to 5000 BC this symbolism correlates with the *movements* we associate with connections needed to start a trading culture in northern Mesopotamia. The three subsequent periods of around 2000 years spanning c. 5000 BC to c. AD 1000 each shows that the symbolism of Gemini is also valid for making the connections necessary for emotional, mental and spiritual intelligence to emerge. The idea of using the symbolism of Gemini for four periods, each of about 2000 years, introduces the concept that the four humanistic phases of Gemini created a continuum spanning around 8000 years.

We know that the physical evidence of Gemini symbolism being expressed in human physical activities is around 1000

---

[6] Guirand, F. *New Larousse Encyclopaedia of Mythology*, p.268.

years before the astronomical Age of Gemini commenced.

The symbolism of Gemini expressed over a period of some 8000 years appears to cause a problem with our model. However, we will see that the new emotional, mental and spiritual connections, made after c. 5000 BC, are more fully described by the physical qualities symbolised by the constellation of the three astronomical Ages of Taurus, Aries and Pisces that followed Gemini.

2. Astronomical Age of Taurus

The symbol of this constellation is the bull and this animal's fertility provides a vital theme to associate with the human condition. The objective is to see if there is physical evidence for the emergence of new aspects of emotional intelligence occurring some 1000 years before the Age of Taurus 4079 - 1929 BC.

2.1. The symbolism of Taurus within the Earth/Sun cycle.

An example from nature during the month of Taurus in the northern hemisphere is late spring's vibrant qualities displayed by the rapid growth of plants. This is illustrated by animals grazing on the new growth of lush grass to provide a supply of rich nourishing milk for the newborn. The abundance of long grasses and new leaves provides the protection many young creatures need as they scurry away from hungry predators. To focus on the growing crops may miss the underlying energies. This error can arise because the green growth of spring is the result of the accretion of resources rather than the intrinsic energies themselves. It is their effect that is so obvious rather than their cause, which lies within the soil, its fertility and structure. For these to be potent the vegetation that supported the life of plants from earlier cycles has had to rot away and recycle nutrients that renew the soil's fertility. This natural alchemy, bringing the gift of life from death, begins in earnest when the Sun is in Scorpio and aligned with the eye of the Taurus bull.

2.2. The star lore of the constellation of Taurus.

The symbolism of the bull was related to particular stars by the ancient astronomer-priests of Mesopotamia. They identified the distinctive features of the constellation of Taurus and associated it with the head and forequarters of the bull. Aldebaran, the 15th brightest star of night sky, marks the eye of the bull. It appears to lie among the Hyades but is much closer to the Earth at a distance of only 68 light-years. The Hyades include over 200 stars and is located along a line between the Pleiades and Orion's Belt. It is the closest star cluster to Earth being 150 light years away. Its V-shape forms most of the bull's face. Two other stars, Zeta Taurus and Beta Taurus, show the bull's horns.

The Pleiades are one the easiest features of this constellation to identify since they form a dense and radiant group of stars, just northwest of the prominent three central stars forming the Belt of Orion in Gemini. At a distance of about 400 light years, they are the most distant stars of the constellation of Taurus. Often known as the Seven Sisters, they form a cluster of over 300 other stars of which Alcyone is the brightest. In Greek mythology, the Pleiades were the daughters of Atlas and the sea nymph Pleione, the 'sailing queen.' Sailors of ancient Greece were advised only to sail when the Pleiades were visible. As these stars vary in their distance from the Earth, we need to recognise the different depths of the symbolism described above.

The star Aldebaran defined the central position of the constellation of Taurus in the Babylonian zodiac. When it is rising over the horizon, another star, Antares, is setting. This phenomenon neatly divides the zodiac in two and creates an accurate natural reference axis.

The symbolism of Aldebaran cannot be considered in isolation since its position and meaning in the heavens must recognise that the star Antares is practically exactly opposite in the constellation of Scorpio. As one of the Royal Stars of Persia, Aldebaran was understood to offer the possibility of personal glory, success or happiness. This becomes more likely if new resources are tapped: in a farming context, this symbolises the need for the renewal of the soil's fertility if good crops have to be grown.

2.3. Historical examples of the symbolism of astronomical Age of Taurus (4079-1929 BC) occurred 1000 years earlier.

To understand the major themes relating to the 2000 year period after c. 7000 BC, we will first consider what was happening in Mesopotamia after 5000 BC. The physical expansion of farming demanded irrigation to release the fertility of this land between the rivers. This soil, with a good structure and high fertility, was a huge physical *resource* for growing surplus crops to trade.

We will next consider the significance of the bull within the ancient Egyptian culture of the Nile valley. The archaeological discoveries made between the Nile and the Red Sea, in the now inhospitable Eastern Desert, describe an ancient dependence on hunting and herding.[7] The domestication of cattle in Africa occurred c. 9000 BC from a different genetic strain to those from western Asia.[8] Toby Wilkinson writes about the later introduction of agriculture and the established cultural importance of bulls within the Egyptian Badarian period, 5000 – 4000 BC. The successful owners of large herds of cattle achieved a high cultural status. The Bull was a physical *resource, valued* for its fertility and became a symbol for wealth and power. The burial grounds reserved for the ruling elite included the graves of cattle.

Another animal, the horse, also came to be *valued* as a *resource* for travelling faster over land. Larger herds of horses could be more easily managed. This practice spread eastwards from the Danube valley and the Carpathian Mountains. Evidence of the wear on the teeth of horses is used to estimate

---

[7]   Toby Wilkinson, *Genesis of the Pharaohs* by p. 101.
[8]   Steven Mithen, *After the Ice*, p.496.

the ages when horses were first ridden with the use of a bit. 'Riding began in the Pontic-Caspian steppes before 3700 BC, or before the Botai-Tersek culture appeared in the Kazakh steppes. It may well have started before 4200 BC'.[9] It is possible that the art of horse riding was introduced some millennia before the Botia culture of northern Kazakhstan established its 99.9% dependence on horses for food, 3700 – 3000 BC.[10] An earlier shaman could have taught about *intuitive* connections with the beast to enable bonds of security, love and play to develop between rider and horse and introduce a motivation, which did not demand the use of a bit. When the horse was used as a source of power, the bit and harness were necessary.

A feature of the mind is that it can benefit from *intuitive* understandings. The potential benefits available from resources found in the natural environment depended on those individuals able to appreciate what was possible, and to lead by their own example.

Another aspect of life where making connections influenced change took place in cultures existing side by side. The expansion of language included cases where communications between different cultures were enriched as words for better farming practices spread to became part of normal life. People from different cultures emotionally engaged and communicated with the new words, which were assimilated because of their usefulness. Historical linguistics provides a tool for us to identify this spread of knowledge. David Anthony describes the origin of this use of linguistics by introducing the work of German Romantics in the 1780s. Herder and von Humbolt established the idea of the role of language with a sense of national identity. The search for a mother tongue has established that the Proto-Indo-European language was the seed of Western civilisation. The homeland of this language is argued to have been between the Caucasus and Ural Mountains

---

[9] David Anthony, *The Horse, the Wheel, and Language*, Ibid, p.221.
[10] Ibid, p.217.

in the region identified as the Pontic-Caspian. The banning of the use of the mother tongue by invading armies is used as a tool to undermine a sense of national identity. The above descriptions illustrate how language plays a vital part in making the *emotional connections* needed for a culture's integrity.

3. Astronomical Age of Aries

The symbol of the ram's horns provides an image of strength and *action*. These characteristics have rich potential for describing overt behaviour. The objective is to see if the physical evidence of the emergence of new aspects of mental intelligence, 1000 years before the Age of Aries, 1929 BC - AD221, is described by this constellation's symbolism.

3.1. We begin by considering the symbolism of Aries within the Earth/Sun annual cycle.

From late March to late April in the northern hemisphere the natural changes in the environment are described by the symbolism of the constellation of Aries. This time of the year is three months past the greatest darkness experienced at the winter solstice and its minimal conditions to support normal life. As the Earth orbits around the Sun, the tilt of the Earth's axis towards the Sun gradually increases the strength of sunlight received by the soil and it warms more quickly. The hours of darkness no longer threaten death. At the spring equinox, the dominance of darkness is overcome and the annual cycle of growth is seen to begin. Signs of initiative being taken in the natural environment are visible. Each germinating seed expressing the will to grow begins to manifest the miracle of life it contains. Here we have the idea of natural sacrifice being part of life; the seed is sacrificed in trust that its life will create more seeds. Buds on tree branches begin to swell and burst into the scented blossom that later attracts the pollinating insects. New shoots thrust through the soil's surface in search of sunlight, and risk being nipped by frosts and hungry beaks.

3.2. The star lore and mythological symbolism of the constellation of Aries.

Although the constellation of Aries has only a few stars, each of modest size, its residents are of significance because they were present at the spring equinox long ago. At this equinox c. 500 BC, the Sun's rising energies are demanding expression as new life begins to dawn. Brady writing describes the symbolism of the largest star in this group, Hamal 'as a forceful star of action and independence'.[11] The sacred ram of Thebes was not 'the annual sacrifice to Amon, but as the god himself, whose identity with the beast is plainly shown by the annual custom of clothing his image in the skin of the slain ram'.[12]

3.3. Historical examples of the symbolism of Aries being expressed as an aspect of emerging mental intelligence.

The symbolism of the ram to describe the archetypal *action* needed to start new cycles has several historical contexts. This animal's symbolism has both royal and spiritual associations within Egyptian culture. The head of a ram with curled horns was at times used on the figure of the deity Amon, king of the gods. Thoth, the eldest son of Ra and the kingdom's sacred scribe, decreed that all kings should come with offerings to the 'living ram' the king of all Egyptian animals.[13] The qualities of the ram, considered as vital to describe Egyptian deities tell of the importance inherent in the king's will, his spoken word, and the *courage* and *action* that the regal word demands.

The decisive *actions* taken by Egypt's first kings created and established the world's first nation. The blessings that flowed from this unification included the *action* to use *education* to bring Egyptian tribes together. This *answer* to solve inter-tribal rivalry led to the successful campaigns undertaken by later Egyptian armies to create the first empire in the Near East.

---

[11]   Bernadette Brady, *Brady's Book of Fixed Stars*, p. 228.
[12]   Sir James Frazer, *The Golden Bough*, p. 501.
[13]   *The New Larousse Encyclopaedia of Mythology*, p. 45.

A history of warfare includes the strength of the human will. Strength of *action* and *courage* are fundamental qualities expressed by soldiers in battle. There was no ambiguity when residents of ancient fortified cities saw their enemies' army approaching with a battering ram; their intentions were clear.

4. Astronomical Age of Pisces

The symbol of this constellation is the two fishes. Its image was a symbol for the duality between Man's body and soul swimming in the infinity of cosmic consciousness. New aspects of spiritual *inspiration* emerged over 1000 years before the Age of Pisces, AD 221 - 2371.

4.1. The symbolism of the Earth/Sun annual cycle.

The symbolism of the constellation of Pisces correlates with the month of the year from late February to late March. The harsh weather during this month would wash and blow away the debris from the past. The winter rains lay waiting in the soil denoting both the end of the Sun's annual cycle through the heavens and life's promise of its new beginning. The raw materials for life's energies have been prepared and with spring's promised warmth, the waiting sap will rise to supply nature's need to create its annual display of abundance.

4.2. The constellation's star lore and mythology.

The faint stars of this constellation create a large V in the heavens with the largest star, Al Rescha, at its apex. This star represents a knot in a cord depicted by the two lines of stars that represent the two fish of Pisces, joined to each other but swimming in opposite directions. This knot between the fish represents 'the joining of different concepts to create wisdom and understanding'.[14]

The fish were used as a symbol of feminine wisdom in many cultures. The Chinese Great goddess Kyan-yin was often

---

[14]   Bernadette Brady, *Brady's Book of Fixed Stars*, p. 313.

depicted as a fish goddess, for the Greeks, Aphrodite Salacia was a fish goddess.[15]   As water dwelling creatures, their element represents the emotional life of divine feelings. In the world of the individual, personal feelings are vital parts of each person's life (even when denied). With the two fishes of Pisces, we have these two dimensions of the astral realm, which can become very slippery fish. Fishermen know all about such matters. The Biblical account of Jesus, as the fisher of men, understood his disciples' feelings better than they did themselves. The verses describing Jesus walking on water to calm the storm, illustrate the capacity to rise above dimensions of experience that engulf mere mortals.

4.3. We will now consider historical examples of the symbolism of Pisces expressed through the connections that fostered new aspects of emerging spiritual *inspiration*.

The 2000 year period commencing more than 1000 years before the astronomical Age of Pisces (the Fishes) AD 221 – 2371 includes the example of the Christian use of the fish as a symbol for their mutual recognition: the Catacombs of Rome abound with this imagery.

Much earlier evidence of *inspiration* comes from the time when monotheism was beginning and Moses was guided by his God to lead the Hebrew slaves out of Egypt. The Jewish holy books were written close to the time when the Vedic scriptures began to provide a spiritual inspiration for the Hindu religion.

Other spiritual themes were gathering strength during this period to contribute their global philosophies. The pre-Socratic rational philosophies introduced by Thales were challenged by the studies of the religious mystic Pythagoras, c. 582 - 507 BC. In his mystery school, the use of numbers contributed towards their understanding of life's mystical dimensions.[16] His quest included the marriage of rational and mystical studies. This

---

[15]   Ibid, p. 311.
[16]   Colin Wilson, *The Occult*, p. 251.

created a cultural tension between these separated spheres of enquiry. This unease later became a divorce when the empirical study of the world advocated by Aristotle was favoured over the metaphysical studies of Plato and the experience of divinity Socrates found within. Later, in Europe during the Dark Ages, the study of the duality between life's physical and spiritual dimensions was frozen. Secular pagan culture became illegal as institutionalised Christianity created traditions of fear and sin to discourage *intuitive* spiritual explorations beyond its doctrines. Islamic culture embraced the achievements of the Greek mind. The Pythagorean quest for truth was not directly resumed in Europe for some 2,000 years. As the Renaissance was ending, the contribution of Copernicus to the Scientific Revolution proved that the astronomical facts he discovered contradicted the 'truth' as decided by the Church.

Concurrent with the work of Pythagoras, other philosophies were born in other regions of the world with very different ways of exploring life's physical/spiritual duality. This 'coincidence' pinpoints a seeding period from which has grown numerous ideas that have shaped the emergence of global cultures. Each has their respective skilled uses of the mind and continues to test human abilities in their different ways of expressing cultural excellence.

The second philosophy we will look at was originated by Gautama Buddha, believed by many to have been born in India c. 563 BC; this teaches that all suffering is created by the desires of the mind. [17] The spiritual practice of observing his Four Noble Truths and following the Eightfold Path leads to the *answer* to the causes of suffering. The condition of non-attachment allows an escape from the continual wheel of rebirth. His teachings added new dimensions to the Hindu religion that had developed

---

[17]   Sangharakshita, *The Three Jewels*, p. 12.

from the 3000 year old Sanskrit writings, *The Upanishads*. This language of Vedic scripture provides a holistic worldview. The translation by Shearer and Russell includes how the Sanskrit word *mushika* is used to symbolise the Self, the unattached pure consciousness.[18] 'To realise the Infinite, whether it be called the Absolute, the Godhead, Brahman or the Self, is the goal of life'.[19] The Buddha taught meditation as the *answer* found on the path to this *intuitive* spiritual experience of the Self's peace and tranquillity experienced with enlightenment. The attachment of the mind to worldly matters dissolves when its inner dimensions are the focus for joy. Buddhism spread widely in eastern Asia and influenced the Chinese culture's beliefs based on the teachings of Confucius and Lao-tzu.

This third philosophy, Confucianism began during an unsettled period of Chinese history with the teachings of Confucius, c. 550-479 BC. In his book *An Introduction to Confucianism*, Xinxhong Yao writes how the work of Confucius was in response to a decline in the feudal order of this time. 'Confucianism is more a tradition generally rooted in Chinese culture and nurtured by Confucius and Confucians rather than a new religion created, or a new value system initiated, by Confucius himself alone'.[20] Before Confucius, Chinese culture's traditions already included the 'six classics', which the scholars of its many schools studied. The conduct of scholars set cultural standards of morality and ethics (*ru*) valued by the early kings and later governments. Shaman and professional gentlemen of virtue contributed in various ways to serve the good of their culture. W. E. Soothill writing about Confucius in his translation of *The Analects of Confucius*, 'He called himself a transmitter, declaring that his philosophy was based on the wisdom of the

---

[18] Shearer and Russell, *The Upanishads*, p. 8.
[19] Ibid, p,11.
[20] Xinzhong Yao, *An Introduction to Confucianism*, p.17.

ancients, whom, as in the *Analects*, he constantly quotes. For him Right Living was the *answer;* essentially the harmonising by the Man of Virtue of his own personality into the social order, and work for its progress and his own'.[21] 'As mentioned above, Lin Xin gave a clear explanation to the origin of *ru.* He traced the origin of *ru* to a government office whose function was to 'assist the ruler to follow the way of the yin-yang and to enlighten [the people with *answers*] by education'.[22] The collection of ancient Chinese writings attributed to Lao-tzu (translation of 'old man of wisdom') are rooted in observations of nature and developed into the practice of Taoism (Daoism). This philosophy's most sacred book the *Tao Te Ching* (Dao De Jing) describes how the *Tao*, or the *Way* is found through meekness and experiencing life's oneness within the Self and to be an innate source of virtue, to be expressed as appropriate. After 200 BC, this philosophy was gradually adapted into the Confucian religion after contact with 'the schools of Legalism, Yin-Yang and the Five Elements, Moism and Daoism'.[23] 'A theological and metaphysical doctrine of interaction between Heaven and humans was established and consequently became the cornerstone of the revived Confucianism'.[24] With these revisions, Confucianism embraced 'the three doctrines' of the Buddha, Confucius and Lao-tzu.

The philosophies originated by Pythagoras, Buddha and Confucius each identified and focused on a sacred aspect of life's mysteries; to understand them, to be part of them and to live in harmony with them. The monotheism of the Abrahamic religions of Judaism, Christianity, and Islam provide a large numbers of believers each with their own one god.

---

[21] W.E.Soothill, The Analects of Confucius, p. xv.
[22] Xinzhong Yao, *An Introduction to Confucianism*, p.18.
[23] Ibid, p.7.
[24] Ibid, p.8.

Review

In this chapter, we have found that over some eight millennia many details of physical, emotional, mental and spiritual qualities of emerging intelligence are described by the symbolism of the astronomical Ages of Gemini, Taurus, Aries and Pisces. The qualities symbolised by these astronomical ages helps us understand more about the growth of intelligence.

The starting dates for these 2150 year astronomical ages, when compared against the evidence of the stages of physical, emotional, mental and spiritual dimensions of emerging intelligence, are about 1000 years later than when these new dimensions of intelligence began to emerge.

This evidence answers the question asked at the beginning of this chapter and tells us that the symbolism of the constellations of the zodiac is a common factor between the annual cycle of the terrestrial seasons and the emergence of new aspects of human consciousness. This correlation is evidence of the principal of correspondence 'as above so below' being expressed. The scientific methodology developed to establish the accurate observations and calculations needed for astronomy to develop, provided the tools needed to study a fourth philosophical way to understand life.

The history of the development of astronomy includes evidence of Babylonian star lists and the names of several constellations, from c. 1100 BC astrolabes.[25] Later evidence from tablets created in 686 BC known as mul.apin list all the names of the constellations of the zodiac.[26] Some two centuries later texts dated to the period 474 BC to 456 BC illustrate the 360° zodiac belt of twelve constellations of unequal lengths. When early scientists decided to change this to make twelve 30° portions of the ecliptic, the twelve signs of the sidereal zodiac were created.[27]

---

[25] Robert Powell. *History of the Zodiac*, p.6.
[26] Ibid, p.6.
[27] Ibid, p.7.

New developments in astronomy became a vital tool for later scholars and astrologers in countless royal courts during the next 2000 years. The Privy Council of Queen Elisabeth 1st provides an example of this by their consultation with John Dee (1527-1608).[28]   He was a scholar of St. John's College, Cambridge, an English professor of mathematics and astronomer, and practised as an astrologer. This culture began its decline with the advent of astronomy separating from astrology with the scientific revolution and the 1660 establishment of the Royal Society.

The power of the intellect was famously used by Isaac Newton (1642-1727). In his later years, his knowledge of precession was expressed in his interest in the astronomical ordering of history as published posthumously in his writing of *The Chronology of Ancient Kingdoms*. [29]

The successful use of the symbolism from the astronomical ages to enrich our understanding of the growth of four qualities of intelligence suggests the possibility that the symbolism derived from the other zodiac constellations may describe other qualities of emerging consciousness. To assess this possibility, we first need to see if we can identify more keywords to complement those already identified. These need to be suitable for describing four dimensions of experience that directly relate to the other qualities of zodiac symbolism.

Over some three thousand years, the art of interpreting the movements of the heavens was studied to gain a better understanding of terrestrial events. The many advisors appointed by the cultural elite of numerous ancient cultures paid much attention to such matters. Much evidence about this interest has survived for us to consider in the next chapter.

---

[28] K.L. Hollihan, *Elisabeth 1st, The People's Queen*, p.40.
[29] Robert Powell, *A History of the Zodiac, Newton's Chronology*, p. 193.

10

A Cycle of Precessional Symbolism

The earlier overview of the emergence of new qualities of intelligence considered their physical, emotional, mental and spiritual dimensions of growth during the last 10,000 years. The qualities expressed in this process are described by part of the symbolism that also describes the annual cycle of the seasons. The zodiac signs of Gemini, Taurus, Aries and Pisces contribute the symbolism that describes these qualities being expressed. This archaeological and historical evidence illustrates how the principle of correspondences considered in Chapter 2, relates part of the annual cycle of seasonal change to our emerging consciousness. These ideas invite us to ask if the precessional cycle of symbolism will describe and tell us when other qualities of consciousness began to emerge.

To explore this possibility we need to expand the lists of physical, emotional, mental and spiritual keywords, derived from the four signs of the zodiac considered in Chapter 9. To describe the dynamics of a complete cycle we will consider all twelve zodiac signs to identify four keywords for each sign.

To do this we firstly identify the monthly physical changes that take place as described by the symbolism of the twelve qualities of the zodiac. The idea is that each of these qualities of change is complemented by keywords identifying an emotional, mental and spiritual correspondence. Our experience of musical octaves enables us to relate these to the emotional keywords. The mental keywords introduce another of life's natural languages. The spiritual keywords describe this cyclical level.

The customary spring starting point of the annual cycle of

the seasons is provided by the sidereal sign of Aries since the Sun was passing through this region of the heavens when this zodiac first came into use c. 500 BC. Ancient seers projected the imagery of the Ram onto this group of stars. We will look to see how the changes that take place in the natural environment during this month symbolise the Egyptian understanding of the divine will of the Aries Ram in action, as we considered in Chapter 9.

After the spring equinox, the increasing daylight reaches the point when the days are longer than the nights. The warming environment aids the strength of the physical *action* as young plants thrust their tender new shoots through the soil and tree buds start to swell. However, the environment can easily be cold enough for frosts to stunt this return of new life into the natural environment. When we think of human *action* to introduce recognised and needed opportunities for change, it is usual for such ventures to be opposed, even if a dwindling minority is sometimes strong enough to resist change. This threat requires us to consider other energies that support physical *action*. We can appreciate how the strength of emotional, mental and spiritual octaves of Aries describe the emotional *courage*, mental *initiative* and spiritual *will* necessary to accomplish the work to be done to ensure life will be lived, whatever the cost.

The physical signs of the symbolism of the sidereal sign of Taurus are around us when the month of May is only about ten days away. Then the fertility and condition of the soil are the resources vital to sustain the new shoots of spring seedlings. When these become flourishing plants, they then become the *resources* to later harvest.

A person's attitude to their physical *resources* often gives us clues about their family's emotional *values*. What people want often changes with their circumstances? Mental opportunities lead to finding different *answers* to problems. New spiritual *aspirations* may introduce realistic expectations.

When June arrives, the Sun has already moved a quarter of the way through the sign of Gemini. The qualities of this symbolism were explored in depth in Chapter 7 when we sought descriptions of the mind learning to make the new connections necessary for intelligence to expand. The theme of Gemini symbolising new *connections* being made in the natural world at this time of the year is provided by the newborn of countless species learning where to find their food.

The survival of many new generations is at stake each spring. Risks abound in most human activities. When a person makes a new physical *connection,* some caution may be exercised as the other octaves of Gemini may intelligently contribute emotional *intuition* about the experience ahead. This may be a good guide as to whether to proceed or not. After the incident has passed, memory records the facts as another experience of *education* takes place. Gemini's spiritual *inspiration* is often at work when introducing the learning experiences required. Each of these four dimensions of experience depends on its own type of connections being made.

The Sun's movement from the sign of Gemini into the sign of Cancer celebrates its cycle's peak at the Summer Solstice. The Sun's maximum noon altitude is held for three days. The seasons change and the next stage of growth is introduced. The symbolism of Cancer describes how love swells the hard new fruits with the sap that *secures* the goodness their crop provides.

When we think of people being physically *secure* this can be the active result of emotional *nurturing.* This emotional dimension of Cancer includes *learning* about the feelings associated with facts, of appreciating the emotional charge information may carry. At the spiritual level, Cancer symbolises the *blessings* love provides within a family or its garden.

During the month of Leo, the environment is at its hottest,

ideal conditions for the ripening of grain and fruit. During August, we appreciate the radiance at the heart of nature's creative power. Individual plants are known by the *identity* their species radiate. Each unique physical form includes the particular characteristics of their fruit and seeds.

People skilled in this botanical knowledge make valued contributions when the potential benefits of plants are harnessed to serve their community. Other people with different creative skills put their heart into many tasks. Butchers, bakers and carpenters have skills attributed to them; these distinguish them from others. With *identity* as a physical expression of Leo, this quality expresses emotional *confidence,* mental *centeredness* and spiritual *self-awareness.*

During the month between late August and late September, the maturing of grain and seeds is described by the symbolism of Virgo. The sensitivity of physical matter to express perfection is illustrated as crops ripen and their harvesting begins. Gathering the fruits of the Earth brings well-deserved celebrations. The goodness of each crop is anticipated and appreciated during the coming months of winter. One theme of life's purposes is *served* with the season's surplus crops harvested and stored.

With perfection as Virgo's objective, the farmer's plants serve another purpose as they mirror their grower's *care* and *discrimination.* The results achieved tell us much about the *purity* of this motivation and purposes served. A way we may find answers to these questions reveals the many complexities the modern world imposes on rural traditions. The symbolism of the next sign of the zodiac, Libra, explains some of these challenges.

The second half of the year begins with the autumn equinox of late September. At the start of the month of Libra, the hours of daylight and nighttime are in balance. Those of darkness soon lengthen and bring the first frosts. With the growing season ended and colder months ahead, life's emphasis changes its focus from individual species to the greater complexities of the environment's capacity to support all species.

The countless shades of autumnal foliage bring spectacular beauty to many woodland vistas. The annual bounty of ripe grain, fruit and seeds provides the food needed to keep the creatures dependent on them alive. The animal, bird and insect residents together with annual visitors are selective when they pick and choose what to eat. Food from each new source is assessed before being eaten. In many ways, purposeful evaluations take place. Numerous examples exist of plant species being dependent on the animals that rely on them for food. Direct *inter-dependence* exists when another purpose is served. Undigested seeds from eaten fruit are spread around the environment in excreta to foster the spread of species. The availability of foods within an environment determines which creatures will find it agreeable to them. The transition from the growth of food to the implications it has for the wider environment introduces a change of focus. The agenda now is the purpose served within life's greater complexity.

The symbolism of Libra is socially expressed as individuals enjoy contributing to the quality of life of others. The theme of relating is of primary concern in many business ventures too. Often the individual realises that their own efforts are more productive when engaged with those of others. An example of this is illustrated in the production of food for our global population. When staples are traded in numerous markets across the world, prices are not determined by the needs of the hungry. The laws of supply and demand ensure that the needs and wants of the wealthy are satisfied first. At each stage of the journey from field to table, the many *evaluations* taking place illustrate how people *inter-depend.* The Libra mental a*greement* between supply and demand can be distorted by greed and power. Friendships and *goodwill* may be at risk when the individual's abilities to contribute towards the needs of their local community have to compete with vested interests.

The spectacular colour of autumn leaves presents its seasonal finality before their death shrouds blanket the ground. Their disintegration begins as fungi, worms and microbes render down old leaves and broken stalks. Scorpio symbolises this power to reclaim the goodness from those parts of nature that supported its greater expression. As they rot, the elements from spent vegetation are *released* and *transform* the soil with enriching nutrients. These will be *shared* with other species when they are again *reborn* as part of life's display of creative power. With autumn's increasing hours of darkness and intensifying cold, the barren landscapes hide the preparations taking place so seeds may soon spring into life.

The powers expressed in the ways life and death follow each other are described by the symbolism of the constellation of Scorpio. Its many hidden mysteries interweave the secrets of sex, death and rebirth. To glimpse the complexities of the many levels involved we will consider how several creatures help describe some human experiences relating to Scorpio's qualities. We will see how the above four keywords help us understand the related spectrum of experiences.

The sting of the poisonous scorpion is a common accident that symbolised physical death. Fixed by fear few contemplate the *release* of spirit from the body, of flesh being *shared* between the creatures in the grave. Fear of this abyss may echo within when less final deaths punctuate an autobiography.

In ancient Egyptian cultures, initiations took place when aspects of ego's personal self, which inhibited the soul's advance, are *released*. Many such deaths allowed the true individual to be reborn, to emerge from the pyramid *transformed* to *share* experiences of life from a higher plane of awareness. With this condition, the symbolism of the eagle becomes appropriate. When the earth-bound power of Scorpio is redeemed, the rebirth of the spiritual self may begin. The soaring eagle lives in two worlds, its vision across the lands below symbolises the free spirit's intelligence to see with great clarity. The eagle's penetrating eyes pinpoint vulnerabilities: the ego may die to be reborn with the intent to identify the common ground enemies *share*.

When personal motivation is directed towards this spiritual level, insight and compassion can replace the dark and selfish abuse of power. Scorpio's energies are thus *transformed;* the symbolism of the dove illustrates the self-discipline necessary to achieve the potential *rebirth* possible. Scorpio's use of the extremes of emotional power are symbolised by the serpent's association with the subjective judgements between healing or destruction, light or darkness, good and evil. The changes taking place between each pair are symbolised by the serpent shedding its skin so that the wisdom of the earth may influence and realise the potential developments unfolding during creative cycles.

Another example of Scorpio *sharing* power is illustrated within a continuum of generations. Each family maintains the common evidence of the immortality of physical traits when a child resembles a parent. This suggests emotional, mental and spiritual themes are similarly interwoven within successive generations. These include opportunities for *rebirth* after the 'death' of 'the sins' of the father or mother.

For another example of Scorpio's symbolism we only have to refer to the practice followed after a person dies and is *released* from their physical life. Their last will and testament often concentrates the minds of those concerned with the deceased's final wishes. When tangible assets are accepted by a beneficiary, they may emotionally *share* the wishes of the deceased. During the transfer of assets, ownership is mentally *transformed.* The responsibility for the use of the bequest is spiritually *reborn* within the new owner.

During the third month of autumn, the botanical stillness in the land is only disturbed by chilling winds. This disguises the qualities of the symbolism of Sagittarius, which accurately describes the theme of expansion being expressed. The life force held in scattered seeds is carried by the autumn winds and rain into the wider world to contribute as required. This brings *opportunities* for a species to grow in new locations. At each place a seed settles, the growing conditions, whatever they may be, must be accepted. This then becomes an *honest* test of the true adaptability of the species. If a seed germinates, it may maintain or extend the presence of a species in a region. If no seeds from a parent plant take root, the species may become extinct and eventually be replaced by a variety more suited to local conditions. Such changes in the natural environment may be seen as an expression of nature's wisdom at work.

The true adaptability of people comes to the fore during this month, the duality between the individual's lower and higher mind is exercised. The ability to identify working principles and handle abstract ideas is useful for organising human affairs and is only limited by people's *belief* in what is possible. When physical *opportunities* arise for an individual, their *honest* emotions reveal their ability to *trust* and be guided by their own *truth*. New encounters may bring experiences that release limiting *beliefs* and expand wisdom.

The winter solstice defines the beginning of the six-month journey of the Sun from its lowest to its highest noontime position. As winter becomes established in the month of Capricorn, the natural world seems dormant. Underground the Earth's core radiates heat to support life beneath the snow. Oceans, mountain ranges, forests and deserts maintain the climate *established* in each region and determine the conditions in which practical solutions for living have to be found. Whatever physical aspect of life has lasted long enough eventually becomes physically *established*; it has passed the test of time and become worthy of some degree of respect for the control exercised. This theme *authorises* its purpose through the mental *knowledge* of the part of life it is. When regard becomes reverence, only then may this control be deemed *sacrosanct.*

Once winter is established, what purpose is served when the month of Aquarius holds the landscape in its icy grip? From late January to late February a sense of detachment may arise when acknowledging the fate that some plants and animals have died while others have the capacity to survive the year's harshest weather. Such times illustrate one of nature's many laws. Various species form diverse *groups* that somehow meet each other's needs and bear witness to the *intelligence* within nature that is capable of nurturing the expression of each species of life. Beyond the laws understood by science, environments are reformed as climates change and species spread. When reform involves the creation of new species, a higher form of intelligence is intrinsic to nature's creative genius. Such displays of life can only be accepted and marvelled at for their wisdom.

Within the many fields of human endeavour, reforms have taken place to raise the quality of life for countless millions. This possibility is alive in different cultures when *groups* of *friends* share *intelligence* able to realise the expression of common human *ideals.*

The signs of new beginnings in the cycle of the seasons come as snowdrops flower and glisten in the sun. When the extremes of cold during February have yet to pass, the month of Pisces has just commenced. The lengthening hours of daylight bring an *acceptance* that each species has its purpose when alive with life, having eluded the death its weaker members met when life slowed almost to a standstill. In fact, the natural world is in a better condition now that the promise of the spring is close. Sap awaits the time plants can *accept* its nourishment and bring new life into the environment.

At this time of the year, we can *accept* that we have nearly completed another cycle of the seasons and have *compassion* for others less able to face life's trials. As we wait for signs of spring to strengthen we can *contemplate* that just as new life is about to be born out of winter's desolation, we too are not separated from the guides to our fulfilment in the year ahead. If inclined, we may also hope that there is some plan perfectly unfolding according to the personal needs of each individual's journey, and the quality of our individual efforts is not insignificant within the collective *unity* of human effort.

This brief journey through the months of the year illustrates how life's energies display the principle of correspondence; life's energy scales resonating throughout nature. Human physical, emotional, mental and spiritual levels of consciousness are just one part of this pattern. We know the signs of the zodiac introduce themes that describe the sets of keywords following the order of the familiar seasonal cycle. Later we will see if these keywords describe the continuum of our emerging consciousness during precessional cycles. The strange thing is that this cycle runs in the opposite direction to that which we expect. This reversal of the signs of the zodiac and their keywords is shown in Table 10 below together with their respective themes. Alongside these are the keywords for the dimensions of physical, emotional, mental and spiritual experience.

The qualities of each sidereal zodiac sign symbolise the physical, emotional, mental and spiritual keywords to be used to expand our model for the growth of intelligence. The movement of precession provides a timetable to be used with these keywords to see if they correlate with the positive growth of the qualities of consciousness that have been discovered, well studied and known to have taken place.

Table 10

Keyword sets for dimensional increments

Precessional order of each sidereal sign with its theme	Physical phase of 2150 years	Emotional phase of 2150 years	Mental phase of 2150 years	Spiritual phase of 2150 years
Scorpio, power.	*release*	*sharing*	*transformation*	*rebirth*
Libra, relating.	*evaluation*	*inter-dependence*	*agreement*	*goodwill*
Virgo, purpose.	*serve*	*care*	*discriminate*	*purify*
Leo, radiance.	*identity*	*confidence*	*centre*	*self-awareness*
Cancer, love.	*security*	*nurturing*	*learning*	*blessings*
Gemini, connections.	*movement*	*intuition*	*education*	*inspiration*
Taurus, fertility.	*resource*	*value*	*answer*	*aspiration*
Aries, strength.	*action*	*courage*	*initiative*	*will*
Pisces, non-separation.	*acceptance*	*compassion*	*contemplate*	*unity*
Aquarius, evolve.	*group*	*friendship*	*intelligence*	*ideals*
Capricorn, control.	*establish*	*authority*	*knowledge*	*sacrosanct*
Sagittarius, expansion.	*opportunity*	*honesty*	*truth*	*belief*

11

A Model for Our Emerging Intelligence

After the last Ice Age ended and the warmer climate had stabilised we found that new aspects of intelligence were beginning to be exercised. Different capabilities developed through physical, emotional, mental and spiritual phases, each lasting about 2000 years. We recognised that soon after farming started the physical *movement* of grain developed trading as a new expression of physical intelligence. This was followed by intelligence learning about emotional, mental and spiritual aspects of life. During each of these three phases physical intelligence developed in different ways.

These consecutive ways of learning about four different physical aspects of life span a period of some 8000 years. Since the astronomical ages follow each other by 2150 years, the idea of relating the 2000 year periods for the growth of intelligence to the astronomical ages suggests a basis to develop a model to help us understand what happened.

We will arrange the ages in a way that shows the four physical stages following each other. When doing this we also need to illustrate that during each age learning opportunities arise involving physical, emotional, mental and spiritual dimensions of experience. We begin by listing the different ages of past millennia against the physical phases that followed each other. The next step is to stagger the start of each sidereal sign's four phases by 2150 years. In this way, each sidereal sign's four-phase 8600 year continuum, overlaps the previous one.

The symbolism of the four recent astronomical ages is taken from Table 10 and shown on Table 11.1 below in a bold font, which is used in two more ways. Secondly, it highlights the continuum of learning intelligence through the four **phases** of the symbolism of Gemini. Thirdly, it shows the continuum of four physical **steps** each concerning a quality from one of the four aligned sidereal signs.

Table 11.1.

The relationships between the growths of intelligence and the astronomical ages.

Sidereal Signs of the Zodiac	The Growth of Intelligence making connections					
	The Astronomical Ages					
	Cancer starting 8334 BC	Gemini starting 6184 BC	Taurus starting 4079 BC	Aries starting 1929 BC	Pisces starting AD 221	Aquarius starting AD 2371
Gemini	physical *movement*	emotional *intuition*	mental *education*	spiritual *inspiration*		
Taurus		physical *resources*	emotional *values*	mental *answers*	spiritual *aspirations*	
Aries			physical *action*	emotional *courage*	mental *initiative*	
Pisces				physical *acceptance*	emotional *compassion*	
Aquarius					physical *group*	

Each sidereal sign's qualities is described by the keywords from Table 10.

The relationships illustrated above in Table 11.1 between astronomical ages and the symbolism of each sidereal signs includes associations that appear unnecessary with regard to aspects of the growth of intelligence considered. This raises the following matter: if the model for the growth of intelligence is to be accurate, other expressions of consciousness need to be explained by the model. We can hypothesise that each sidereal sign of the zodiac contributes its symbolism to describe their related 8600 year period of growth for physical, emotional, mental and spiritual qualities of consciousness.

We recognise that the continuum of expanding physical intelligence provided by archaeological evidence begins about 1000 years before the start of the related astronomical age of Gemini. Table 11.1 needs to be adjusted to correct this 1000 year time lapse. An advantage arising from this realignment is the practical use of humanistic ages. To illustrate this we develop Table 11.1 by sliding the line with the dates of the astronomical ages to the left by some 1000 years. This then aligns astronomical periods with humanistic periods. However, we will not yet call the period commencing 6148 BC the humanistic Age of Gemini. We will first consider an earlier observation of the way developments contributed to our emerging consciousness.

When we looked at the expression of spiritual intelligence in Chapter 6, the energies driving the growth of consciousness included the inspirations of Pythagoras, the Buddha and Confucius. The creation of the three philosophies they initiated, their widespread acceptance and the influence they had in shaping their respective cultures some 2500 years ago is difficult to overestimate. This evidence suggests that the prevailing spiritual quality justifies being used to identify its humanistic age.

After these three philosophies successfully shaped their respective cultures, Table 11.1 shows another spiritual quality was naturally introduced. This change identifies *aspirations* as the spiritual imperative that currently dominates global agendas. This motivation is the fourth phase of the continuum of qualities symbolised by the sidereal sign of Taurus.

The clearest and most pervasive illustration of this motivation concerns the modern thirst for still more power and the creation of different types of groups able to wield their collective influences. Since the Scientific and Industrial Revolutions, the quest for power has dominated global history. Central to this objective is the ownership and exploitation of natural resources and the continuing damage to global ecosystems. These activities amount to the slow industrial murder of the Earth Mother.

We will now assess if the spiritual qualities of earlier periods justify being used to name their own humanistic age. To assess this we will modify and expand Table 11.1 by using the above spiritual evidence of the motivation driving recent humanistic ages. We have used the phase of spiritual intelligence in Table 11.1 to identify the humanistic Age of Gemini (1000 BC-AD 1000). The *aspirations* associated with the humanistic Age of Taurus (AD 1000 BC-AD 3000) concern *group* endeavours. The keywords for each phase of Table 11.2 below come from Table 10.

During the first three humanistic ages shown in Table 11.2 above, we will explore the relationships between the *physical* evidence of change and the evidence that suggests *spiritual* motivations were responsible for creating the changes identified.

From Table 11.2 we have three earlier pairs of spiritual and physical keywords as follows, *purify* and *movement*, *self-awareness* and *resources*, and lastly *blessings* and *action*. For each of these three pairs of keywords in their period we want to see if the spiritual quality supports the known physical changes.

Table 11.2.

Phases of Consciousness during each
proposed Humanistic Age.

Sidereal signs of the Zodiac	Estimated starting dates of proposed Humanistic Ages				
	Virgo 7500BC	Leo 5250 BC	Cancer 3100 BC	Gemini 1000 BC	Taurus AD 1000
Virgo	Spiritual *purify*				
Leo	Mental *centre*	Spiritual *self-awareness*			
Cancer	emotional *nurturing*	mental *learning*	spiritual *blessings*		
Gemini	**physical *movements***	**emotional *intuition***	**mental *education***	**spiritual *inspiration***	
Taurus		physical *resources*	emotional *values*	mental *answers*	Spiritual aspirations
Aries			physical *action*	emotional *courage*	Mental initiative
Pisces				physical *acceptance*	Emotional compassion
Aquarius					Physical group

From c. 7500 BC - 5250 BC when farming began to supplement food supplies from hunting and gathering, food surpluses provided opportunities for trading. This major cultural development resulted in growing populations and depended on the success of farming and traders. At this time, metallurgy was refining its skills to *purify* gold and silver ores. These metals were used to create artefacts that became highly valued as goods for trading. These were superseded by different currencies and coinages. The physical *movement* involved when buying and selling a harvest surplus helped establish wealthy merchants working close to agricultural regions.

From c. 5250 BC - 3100 BC in Mesopotamia, when its fertile rainless plains began to be irrigated, it required leaders to emerge with the *self-awareness* to know they had skills to manage the physical *resources* and develop their 'hydraulic culture'. In this way, their trading culture expanded and eventually led to the invention of the world's first cities.

From c. 3100 BC - 1000 BC the tribes of Egypt were united into one kingdom by kings (initially) *blessed* with the vision of divine guidance that enabled them to rule their land for the benefit of the Egyptian population. The physical *actions* they ordered established the world's first nation.

In each of the three periods considered above, we have seen that spiritual qualities introduced conditions that assisted the new physical changes of the period. From this, we have looked at five cases that justify the choice of identifying each humanistic age by its prevailing spiritual quality.

We recognise that the accuracy of the astronomical ages provided a timetable that was estimated to be consistently about 1,000 years later than each of the four periods of the growth of intelligence described. This estimate was useful in identifying the humanistic ages that provided a better fit with the archaeological and historical evidence. However, the matter of accurately aligning the symbolism of the precessional cycle with this humanistic view of history has to be addressed. This is the subject of the next chapter.

12

Degrees of Accuracy

To aid an understanding of our emerging consciousness, we want to see if we can correlate the different milestones of growing intelligence with the cycle of symbolism of the precessional zodiac. To achieve this alignment of the historical record with the precessional cycle we need to correlate the contributions made by individuals recognised for the new qualities of consciousness they introduced, and which contributed to the continuum of human progress. To assess this, we will first consider the precessional cycle in more detail.

The Penguin Dictionary of Astronomy gives the duration of the precessional cycle as 25,800 years. This period is not constant since very small variations in the speed of precession take place due to gravitational influences. An interesting fact about this cycle's duration is the period for each degree of precessional movement. The cycle time of 25,800 years divided by 360 gives the answer of 71.67 years. This is close to the Psalm 90's biblical estimate of the life span of an individual: three score years and ten.

One method of aligning the historical record with the precessional cycle is based on the philosophical premise that a specific spiritual energy is pre-immanent during each of the three hundred and sixty 71.67 year periods in the precessional cycle; each degree has its symbol. People that align their own energies to that of the cycle may then express these creative spiritual essences. In this way, change contributes to the creation of different cultural advances. A simple way to illustrate this premise is to use a major point in the Earth's annual orbit of the Sun to provide us with the symbolism that correlates with a vital period of history.

The historical evidence we assessed to define the humanistic ages listed on Table 11.2 depended on correlating this evidence with the symbolism of five signs of the sidereal zodiac. Within the year's seasons, the related months concerned run backwards from September, through August, July and June to finish in May. The annual order of these months is reversed to allow for the motion of precession. When we consider these months of summer and ask if there is an outstanding moment, the Summer Solstice immediately comes to mind. Pondering over the many historical events we have looked at, it seems likely that Egypt's long association with its Sun cult could provide the correlation sought.

An additional matter for us to consider concerns the increasing and decreasing elevation of the noon Sun. As the summer solstice is reached, this vertical speed appears to reach zero when no vertical movement is observed. Ancient astronomers recognised the phenomenon of the Sun appearing to remain in its noon elevation for some days before being seen to decline. This means that we have to consider at least a two-day period when the Sun appears to maintain a constant noon elevation either side of the moment the solstice is reached. When we relate this fact to the precessional cycle, we have a near 150 year period when the symbolism of the Sun is likely to be the focus of powerful cultural events.

The Egyptian Sun Cult

We will now assess the possibility of aligning the historical record with the precessional cycle. To do this we will see if we can find correlations between historical events from Egypt's fluctuating Sun cult and symbolism from the precessional zodiac.

We noted in Chapter 5 how the priests of Heliopolis near to the Giza pyramids followed the original Sun cult. This became Egypt's national religion, which united the tribes of the two kingdoms of Egypt into one nation, c. 3200 BC. The priests of Heliopolis pyramids worshipped the Sun as Ra or Re, the 'Lord of the daytime sky', and the radiant source of life sustaining energies. The Sun ruled the Egyptian pagan pantheon. The falcon god Horus was the divine ancestor of the Pharaohs. The king's name included the title 'son of Horus' as a statement of his divine lineage. With this status came the pharaoh's role as High Priest of Ra. Later the Sun became the symbol for the Atum, 'the formless spirit behind all creation'.

For around 2000 years, the interests of Egypt waxed and waned. The pyramid culture of the Egyptian Old Kingdom lasted for several centuries after 2630 BC but by 2151 BC had declined. Another climax in this process was reached during the New Kingdom with the worldly success of Ramesses II, c. 1279–1212 BC. Even his brother Ramesses III could not arrest the national decline that followed. Is there some sort of spiritual peak within this period that correlates with the symbolism of the summer solstice?

To explore this possibility, we need to trace briefly the mixed fortunes of the extensive lineage of Egyptian kings. During the period termed the Old Kingdom, famous for its pyramid culture, kings were known as Pharaohs and accepted as divine beings. This helped to maintain national cohesion. In later centuries, this role gradually took second place to economic and other worldly ambitions. After successive Nile floods failed, the disastrous famine that followed finally ended the period of the Old Kingdom. The rise of local warlords illustrates how fragmented Egypt became. These leaders included those of Asian Hyksos tribes that had settled in northern Egypt. The help the Amun priesthood gave to their kings to remove these tribes was rewarded with greater responsibility for the religious and political leadership of their nation. The name Amun means the hidden one, 'the Sun as it journeys through the dangers of the underworld between sunset and sunrise'.[1] Thebes emerged as the new centre of power and became the spiritual heart of the nation. The New Kingdom that emerged from this period of uncertainty saw a succession of Solar Kings coming to power. They aspired to recapture the divine authority held by the early Horus Kings.

The successes of the Solar Kings reached their peak close to the end of the 18th Dynasty 1567-1320 BC. Amenhotep III, c. 1390-1352 BC was the ruler of Egypt at the peak of its national and international glory and much revered as a sage, magician and holy man.[2] During his reign, he exercised great wisdom as he gradually elevated the role of kingship to exercise new heights of power and authority that surpassed the divine status of the earlier Solar Kings. Moreover, it was during his reign that the visible Sun-disc god the Aten was favoured. Such attention was encouraged by the priests of the temple of Heliopolis near Giza. However, in central Egypt at Karnak, the traditional power base of the Amun cult, its priesthood was not disposed to accept this initiative.

At this time, the priesthood was also challenged to adapt its theology to accommodate cosmopolitan influences reaching Egypt from its new empire. While the Egyptians had exported their traditional religion with their conquests, conflicts were recognised. These concerned their spiritual beliefs about religious universalism versus the particular interests of Egypt. 'If all men were created by god, how could one nation claim the right to impose itself on others'? [3] Some religious flexibility was exercised by the worship of Amun-Re and the Aten alongside the traditional Amun.

It is easy to imagine how these pressures for change did not suit an Amun priesthood comprising of 'hereditary bodies grown

---

[1] Stephen Quirk, Ancient Egyptian Religion, p. 17.

[2] Nicholas Reeves, Akhenaten, Egypt's False Prophet. p 66.

[3] Paul Johnson, The Civilisation of Ancient Egypt, p. 86-9.

rich, worldly and corrupt – thanks to past Pharaoh's generous acknowledgment, in word and deed, of Amun's support in his martial endeavours'.[4] (Earlier Pharaohs successfully expelled the Hyksos invaders from the Nile delta region with the help of the Amun priesthood.)

The events that transpired during the latter years of the reign of Amenhotep III became so out of character with that of earlier or later times it has been impossible for scholars to agree about the significance of the changes made during the reign of Amenhotep IV, c. 1352 - 1334 BC. The book *Akhenaten- Egypt's False Prophet* by Nicholas Reeves is used to highlight some key themes of this period.

At this time, it was customary practice for the throne of a king to be passed to his chosen successor before death. When this happened, a period termed co-regency existed. Some Egyptologists have argued that such an arrangement existed between Amenhotep III and his son Amenhotep IV who later renamed himself Akhenaten.

The unprecedented wealth and worldly power of the kings and priests arose from the resources of the regions surrounding Thebes. The uneasy alliance between the Royal Court and the priesthood deteriorated after the priests' religiously opposed Amenhotep IV after he changed his name to Akhenaten and enforced his own new spiritual practices. These focused on the radiance of the Sun during daytime and the benefits this provided to all life. This new religion was in stark contrast to a part of the religion of the Amon/Amun priesthood centred in the heart of the former Upper Kingdom at Thebes, which prioritises the nighttime *portion* of the Sun's daily cycle. The religious traditions concerning the nightly journey of the Sun were symbolically associated with the underworld, with death and the afterlife.

The diminution of royal patronage to Amun religious

---

[4] Nicholas Reeves, Akhenaten, Egypt's False Prophet. p 33.

traditions became so extreme that Akhenaten made its priests redundant. However, despite the initial successes of the new form of worship, these proved insufficiently durable to create a lasting new order. The Amun priesthood undoubtedly aided this reversal in their fortunes. After retrieving and exceeding their former power, they attempted to eradicate all evidence of the spiritual deviation introduced by the heretic king Akhenaten.

The circumstances that triggered these upheavals concerned the revolutionary beliefs arising from the proclaimed divinity of this young pharaoh's father. As junior ruler, Amenhotep IV (Akhenaten) identified the radiant disc of the Sun as the new and sole official God of the nation. This change was related to the symbolism of the Sun as practised during the period of the Egyptian Old Kingdom over 1,500 years earlier. At that time, the Ra priesthood of Heliopolis worshiped Ra Atum as the supreme God of all creation. All the early pharaohs were known as the sons of Ra.

When Amenhotep IV introduced the practice of using the Sun disc to symbolise his own father's divine solar identity, possibly before his father's death, he was reinterpreting very sacred symbolism. The disc of the radiant Sun was known as the Aten. This name is interpreted as, 'he who is effective on Atum's (God's) behalf.' Shortly after this, Amenhotep IV, (meaning Amun is satisfied), changed his name to Akhenaten (meaning the glory of Aten) and became the High Priest of the Aten. He was solely in charge of the worship of his own father. It was this change of spiritual responsibility that made the Amun priesthood redundant. If this took place before Amenhotep III's death, it constituted a new cult with the worship, at Karnak, of kingship in its own right. If not, this heresy arose after the new king proclaimed his own divinity from the belief that he was the son of the spiritual creative source of life, expressing God's Will on Earth.

It was hardly surprising that these heretical practices were unpopular with the redundant Amun priesthood and those still loyal to traditional beliefs. Thebes was not a satisfactory place for the new pharaoh's palace since this was the home of the former priests. With Egypt's vast power and wealth at his disposal, Akhenaten moved his residence away from the centre of religious criticism to a new Egyptian capital city, which he had built. It was located on a pristine site alongside the Nile known as el-Amarna and named Akhetaten, (Horizon of the Aten). It was the centre of national administration as well as religious worship. It made both Memphis and Thebes redundant to the new direction of Egyptian culture. The time when the Egyptian capital city changed in this way is known as the Amarna period, c. 1352 - 1328 BC.

The vast sunlit courtyard of the Temple of Aten at Akhetaten was designed for large congregations of Nile residents. This contrasted with the dark inner sanctums of Amun Temples reserved only for priests. Radically slaves were included in the celebration of the Aten. The worship that took place in this Temple can be interpreted as the public praise of God. 'Hymns could thus be sung and understood not only by the inhabitants of the Nile valley but also foreigners. All men, they proclaimed, were equally the children of the Aten'.[5] Some scholars believe Akhenaten himself wrote this hymn.

After Akhenaten's death, his nine-year old son Tutankhaten, meaning living image of Aten, came to the throne at el-Amarna. Soon afterwards, his name was changed to Tutankhamun, meaning living image of Amun. This showed the priesthood retrieving its former spiritual power and re-establishing former traditional religious allegiances.

Before the young pharaoh was old enough to reconcile the Amun and Ra priesthoods, he died aged nineteen, c. 1324 BC. The throne passed to his guardian Ay, a military leader of Akhenaten averting the possibility of a return to the loss of

---

[5]   Larousse Encyclopaedia, Egyptian Mythology, p. 32.

national power caused by the former divisions and chaos. El-Amarna was gradually destroyed alongside all its official records. In addition, the Amun traditions were changed to include Ra in their sacred solar symbolism.

Within fifty years, the famous pharaoh Ramesses II began his seventy-year reign apparently taking Egypt to its historical pinnacle of opulence, c. 1279-1213. Early in this period, he oversaw the final destruction of the Temple of Aten. The worship of this new pharaoh and his achievements as a warrior included the building of vast monuments proclaiming his own greatness.

The period from the start of the reign of Amenhotep III (Akhenaten's father), 1390 BC, to the end of the reign of Ramesses II, 1213 BC spans 177 years. During this time, Akhenaten's extensive efforts to establish and maintain a national religion based on the belief and worship of a monotheist God as the divine Father of all creation was symbolised by the radiant disc of the Sun. This growing cultural focus, in the context of the cycle of precession, correlates with the seasonal climax expressed as the Sun reaches the peak of its annual cycle at the summer solstice. When we use this time in the precessional cycle, we have a period in history that allows us to align this cycle with the historical record.

This result has been possible because the model developed to understand our emerging consciousness follows the principle of correspondence in two ways.

1. The annual cycle of the seasons provides us with the symbolism described by the twelve signs of the sidereal zodiac. These qualities of time correspond with the twelve qualities of the humanistic ages within the precessional cycle. In this case, we see cycles within cycles.

2. The twelve qualities identified from the annual cycle of the season each correspond with physical, emotional, mental and

spiritual realms of experience. In this case, we see dimensions or planes of qualities.

To determine an accurate alignment between the humanistic ages and the historical record we need to consider the above 177 year period of Egyptian history in more detail. We will explore for evidence that the changes attributed to one pharaoh clearly expressed the summer solstice symbolism during and after a famous cultural peak. We earlier considered the build-up of the expression of solar symbolism; we will now focus on identifying some form of cultural decline after a peak. This correlates with the solar movement of the Sun beginning to lose its height in the noon sky after the summer solstice.

To check for this, we need to identify a critical time in the reign of Ramesses II when his power began to wane and begin the decline of Egyptian power and influence. For this, we turn to the continuing struggles for power in Egypt. Many believe that this competition for power was exercised after the reluctant appearance of the famous Hebrew patriarch, Moses.

The story of Moses began, according to popular biblical history as recorded in the Book of Exodus, when a royal princess rescued the infant Moses from the Nile. Raised in a highly cultured environment he later killed an Egyptian slave-master for abusing an Israelite slave. After fleeing to the neighbouring land of Midian, Moses became a shepherd where he first encountered the one God who would ultimately use him to free the Hebrew people from Egyptian slavery. This meeting is recorded in the story when Moses saw a desert bush burst into flames and miraculously did not burn. Moses saw God in the fire and heard God call him to demand of the Pharaoh that the Israelites be set free. Moses' initial order to the Pharaoh failed. Angry at such interference, the Pharaoh decreed that the slaves must work harder.

As the Pharaoh would not release the Hebrew slaves, a form of divine punishment was visited upon the Egyptians by God in the form of ten plagues described in the Book of Exodus. A form of competition arose between the two leaders concerning who had the most power. The biblical account describes the exchanges between Moses and the magicians of the Egyptian Royal Court. The Israelite God, through Moses, performed the miracles of the plagues to show Ramesses II that his God was more powerful than the Court magicians were. It is possible that Ramesses II delegated royal responsibility for magic to his son Prince Khaemwaset, the High Priest of Ptah. He was renowned for wisdom and magical skill, a reputation that lasted a thousand years into the Roman era.

Chapters 11 to 14 of Exodus describe how the last plague took the life of the first-born son in each Egyptian family, including the Pharaoh's. Each Hebrew first-born child was protected from this catastrophe. Thanks are still remembered with the Passover celebration. This miracle, attributed to the Hebrew God, was followed by the Pharaoh begging Moses to lead his followers out of Egypt. Later, Rameses II regretting this decision, set out with his army in pursuit, to recapture his slaves. Before the Egyptians could catch them, they entered the reed beds and the waters miraculously divided allowing the released Hebrew slaves to pass through to freedom. This final miracle concluded the show of God's power through Moses leading the Hebrew slaves to freedom.

This focuses on a controversial and pivotal period of Jewish history since there are various arguments concerning the timing of the Exodus of the Jews from Egypt and their arrival at Canaan. The title *Exodus* by Prof. W. Johnston published as one of the Old Testament Guides by the Sheffield Academic Press, clearly details the historical, institutional and literary complexities of this subject. Simplistically there is the 'long' chronology, which times the Exodus at c. 1436 BC and the 'short' biblical chronology, which places the migration between 1304 and 1237 BC.

A recent contribution to the debate concerning the

chronology of Exodus provided in the book *Between Migdol and the Red Sea* by Carl Drew, contains two main themes. Whether the Hebrew Exodus from Egypt really happened, and that beliefs and science can and should be compatible with each other. Drew's date for the Exodus is 1250 BC.

We end this chapter by concluding that the worship of the Aten captures the expression of symbolism of the summer solstice. The insights from this chapter will be used to conclude that the precessional 'summer solstice' is estimated to have been c. 1260 BC, early in the reign of Rameses II. This alignment identifies the period for each precessional degree of Gemini.

Table 12.

Periods of history aligned with the precessional movement through the degrees of Gemini.

Precessional order of the degrees of Gemini	Starting date for nominal periods of 72 years. Exact time used 71.67	Precessional order of the degrees of Gemini	Starting date for nominal periods of 72 years. Exact time used 71.67
30	1,260.0 BC	15	185.0 BC
29	1,188.3	14	113.3
28	1,116.7	13	41.7
27	1,045.0	12	AD 30.0
26	973.3	11	101.7
25	901.7	10	173.3
24	830.0	9	245.0
23	758.3	8	316.7
22	686.7	7	388.3
21	615.0	6	460.0
20	543.3	5	531.7
19	471.7	4	603.3
18	400.0	3	675.0
17	328.3	2	746.7
16	256.7	1	818.3

NB. The Appendix describes the technique used to align individual degrees of Gemini by correlating their own symbolism for 72 year periods with the lives of many significant contributors to the growth of consciousness.

13

Expanding our Model for the Growth of Intelligence

An understanding of the growth of intelligence arose from recognising how it developed through physical, emotional, mental and spiritual phases, each lasting around 2000 years. The progress made during the physical phase was later improved when new emotional aspects enhanced the physical intelligence being expressed. This idea of existing capabilities being extended by contributions from later phases was expressed throughout the millennia. The process of becoming more adept at making connections in all four dimensions extended over 8,600 years. The first of these phases of growing intelligence began around 7000 BC, some 1000 years before the astronomical age of Gemini began, 6,184 BC. The symbolism of this sidereal sign correlates well with the physical movement of surplus foods. This trading culture followed the success of agriculture arising from the warmer climate created as the last Ice Age ended.

The astronomical ages that followed are part of the precessional cycle created as the spring equinox moves backwards through all twelve sidereal signs of the zodiac to create a cycle of 25,800 years. The astronomical Age of Gemini was followed by the three Ages of Taurus, Aries and our current astronomical Age of Pisces. The symbolism for these later ages describes the qualities that enhanced the emotional, mental and spiritual growth of intelligence. These are only a few of the twelve symbols of the zodiac portraying the annual cycle of the seasons. This suggests that the timing and growth of other realms of consciousness are described by precessional cycle symbolism still to be considered.

The alignment made between the precessional cycle of zodiac symbolism and the historical record allows us to introduce the idea that within the zodiac structure of the astronomical ages there are also 2150 year humanistic ages. The evidence presented in this work suggests that an 8,600 year continuum of growth symbolised by each sidereal sign gradually develops through physical, emotional, mental and spiritual dimensions of experience. The times when new artefacts began to be used are physical evidence of when human imagination expressed new spiritual motivations. The 2150 year period when a sidereal sign's quality of spiritual experience is being gained is used to name each humanistic age. At any time, all four dimensions of experience are available for people and each dimension is described by the symbolism of a different sidereal sign. This structure is created by successively overlapping the 8,600 year continuums by 2,150 years. This overlapping creates a structure within the 25,800 cycle in which each sidereal sign of the zodiac has the potential to describe the emergence of its own qualities of consciousness to complement the four dimensions of intelligence already identified.

These dynamics are represented by patterns showing two ways each constellation's symbolised quality contributes to the growth of consciousness. Initially each quality consciously develops through its own 8,600 year continuum's four phases. Secondly, during the last phase of this process its spiritual expression becomes the primary motivation during its own humanistic age. It influences the expression of mental, emotional and physical dimensions of experience according to the quality symbolised by the aligned sidereal sign. This structure was illustrated when inspiration, as the expression of the Gemini spiritual dimension of consciousness motivated the growth of mental answers, emotional courage and physical acceptance.

This evidence that the symbolism of the sidereal sign of Gemini within the precessional cycle correlates with the growth of physical, emotional, mental and spiritual dimensions of human intelligence suggests that the symbolism of other sidereal signs of the zodiac are also of profound significance for human consciousness. We need to assess further the idea that each sidereal sign of the zodiac provides symbolism describing other qualities that have already or are yet to come into consciousness. To begin this assessment, we will look for evidence of a humanistic age before that of Gemini earlier identified and explored. For help with this, Table 11.2 is amended below to create Table 13.

We identify the humanistic Age of Cancer spanning the years from 3410 BC to 1260 BC. This period is the fourth and last phase of the 8600 year continuum of new physical, emotional, mental and spiritual dimensions of experience expressing the symbolism of this sidereal sign. We can expect to find that the spiritual qualities of Cancer provided a major motivating influence during its own humanistic age. During this period, its spiritual growth can be expected to be expressed in mental, emotional and physical dimensions as described by the keywords from the sidereal signs aligned horizontally with these dimensions as shown in Table 13 below. In this way, different qualities of consciousness interweave with each other and manifest as expressions of life appropriate to the time and circumstances. The keywords for other threads of consciousness that interweave are also shown in Table 13 to provide a wider perspective of a portion of the 25,800 year precessional cycle.

The symbolism of Cancer during its own humanistic age will be considered in the context of several aspects of Egyptian history during a continuous period of over some two thousand years. The widely researched culture of ancient Egypt provides us with many opportunities to assess the development of its spiritual, mental, emotional and physical dimensions of experience. This historical record presents us with a time-scale that fits well with the duration of a humanistic age.

An interesting theme contained in the method being used to introduce this philosophy concerns the way each quality of consciousness is introduced and develops. To aid the expression of spiritual qualities, experience is progressively gained as its phases of physical, emotional, mental and spiritual experience build on each other. The order followed is shown horizontally in Table 13, physical security, emotional nurturing, mental learning and finally spiritual blessings. This process lasts 8,600 years.

Table 13

Seven Humanistic Ages with their
Dimensions of Consciousness and Associated Symbolism.

Sidereal Signs of the Zodiac	Starting dates for recent Humanistic Ages						
	Scorpio 12,010	Libra 9860 BC	Virgo 7710 BC	Leo 5560 BC	Cancer 3410 BC	Gemini 1260 BC	Taurus AD 890
Virgo purpose	Emotional care	mental discrimination	spiritual *purify*				
Leo radiance	physical identity	emotional confidence	mental *centre*	spiritual *self-awareness*			
Cancer love		physical *security*	emotional *nurturing*	mental *learning*	spiritual *blessings*		
Gemini connections			physical *movement*	emotional *intuition*	mental *education*	spiritual *inspiration*	
Taurus fertility				physical *resources*	emotional *values*	mental *answers*	spiritual *aspirations*
Aries strength					physical *action*	emotional *courage*	mental *initiative*
Pisces non-separation						physical *acceptance*	emotional *compassion*
Aquarius reform							physical *group*

When we consider the physical, emotional, mental and spiritual dimensions being expressed during each humanistic age they are listed vertically on Table 13. During the humanistic Age of Cancer, when we consider the order of the physical, emotional, mental and spiritual dimensions this sequence follows that of the annual cycle of the seasons, i.e. Aries, Taurus, Gemini and Cancer.

Our test for the proposed humanistic Age of Cancer begins by following the method used in Chapter 9 where we looked at the symbolism of the sidereal sign of Gemini to see how it contributed to our understanding of the growth of physical, emotional, mental and spiritual intelligence. To use this method for the humanistic Age of Cancer we will assess the following three points.

1. The natural dynamics expressed as signs that this sidereal sign's symbolism manifest in the environment during the month from late June to late July, in the northern hemisphere.

2. The star lore of the sidereal sign.

3. The mythology and historical examples expressing the symbolism being examined.

The first examples will come from the periods shown in Table 14.1 aligned horizontally from the name and theme of Cancer. After assessing this continuum of symbolism from 9860 BC to 3410 BC, we reach the beginning of the humanistic Age of Cancer when spiritual blessing become the focus of growth. When we consider this humanistic age we may expect to find evidence of qualities of consciousness, which developed earlier, influencing the changes taking place during the period being considered.

We will now consider the sidereal sign of Cancer by looking at the above three points.

1.   The symbolism of Cancer expressed within the Earth/Sun annual cycle.

In the natural environment from late June to late July, during the month of Cancer we need to observe characteristics that identify the spiritual keyword that captures the quality of the birth and initial development of the Egyptian nation.

The unique characteristic of the Sun's motion in our sky just after the summer solstice heralds the changes in nature that provide our first clue. After six months of increasing daylight, the Sun is just past its highest place in the sky. The symbolism of the sidereal sign of Cancer begins to be seen as the Sun's elevation in the sky begins to decline. Life has been renewed after winter's desolation. The agenda now changes to bring forth life's fruits.

The seasons crops have passed the time when they are most vulnerable, have grown tall and flowered during springtime's rush. Each plant's characteristics have passed the tests of time. Now, well rooted, the rising sap swells the fruit of pollinated flowers. Each nourishing fruit soon acts as a womb that nurtures its seed for a later generation. For these blessings to be realised, farmers and gardeners know that each shower of rain brings another *blessing* for thirsty plants. All the work invested in the preparation and planting is vulnerable when droughts threaten.

2. The star lore of the sidereal sign of Cancer.

In the northern hemisphere, the stars of Cancer can be seen overhead during early spring evenings. The particular aspects of Life the ancient priests associated with the sidereal sign of Cancer have to be considered in the context of the seasonal environment as the Sun enters this sidereal sign. To provide us with some ideas to consider we will look at the two main features of this sidereal sign. These are the alpha star Acubens, and Praesepe (the Manger or Crib) with the two nearby stars Asellus Borealis and Asellus Australis, which are close to the sidereal sign's centre, depicting the donkeys of the Manger. This open cluster of stars is known by some as the Beehive, a name perhaps due to the indistinct haze created by the 'swarm' of some 500 faint stars forming the nebula.

3. The mythology and historical examples that express Cancer's symbolism.

a)   A clue for the physical keyword able to express the qualities of this sidereal sign is given by the creature used as the symbol for this region of the heavens, the crab. Its delicate flesh is armoured by protected its shell armour. This protection introduces the idea of how *security* is necessary to create a safe environment for a home. Living in a safe home is the preferred choice of all mothers when *nurturing* young children. This environment, when free of dangers allows the early lessons of life to be *learned* naturally. The joy each baby brings to its family is a *blessing* each generation gives and receives.

These keywords represent the symbolism of the sidereal sign of Cancer we seek to identify being expressed in the new qualities of consciousness emerging as cultural activities change during the four phases of this sidereal sign's 8600 year continuum. The model we are evaluating requires that during each phase of this continuum the new qualities introduced during earlier 2150 year periods are increasingly expressed and refined.

The mythology associated with the sidereal sign of Cancer varies between different cultures. While western traditions link it with the crab, in Babylonia c. 4000 BC the tortoise was chosen to illustrate its symbolism while the Egyptian's chose the Scarab beetle. The choice of three heavily armoured creatures chosen by different cultures to carry the symbolism suggests an archetype where natural safety depends on the security provided by an external skeletal structure.

According to Vehlow, the Chinese gave this group of stars the name 'the spirits of the ancestors'.[1] This describes the continuity of the generations and all that this has established for the families' living representatives. 'The Egyptians, however, linked the sidereal sign to the Scarab beetle, a symbol of immortality, rising from the dead and settling 'upon the empty boat of Ra', a place where the sun god was born again'.[2] For additional insights into ancient Egyptian beliefs associated with this sidereal sign, we will consider the significance of this beetle as one creature of the Ennead of Heliopolis and the family of Osiris. 'For the Heliopolitans, Osiris represented the rising Sun which, like the Scarab, emerges from its own substance and is reborn of itself.' 'Khepri was the god of the transformations which life, forever renewing itself, manifests' signifies at the same time 'Scarab' and

---

[1] Ebertin- Hoffman, Fixed Stars and their Interpretation, p. 45.

[2] Bernadette Brady, *Brady's Book of Fixed Stars*, p. 253.

'he who becomes'.[3]

In a Christian context, the stars of Praesepe were recognised as the Manger or Crib. 'This is a faint nebula which was seen in ancient times as the cradle of life, the point where life emerged'.[4] This nebula was known as the Beehive with its queen, a symbol of the Goddess, at the heart of its existence. However, the gender of the Egyptian monarch was seldom a queen. Despite this, we can see his swarm as an analogy for the loyal service Egyptian workers gave their pharaoh and the royal authority he dedicated for his nation's success.

b). Historical examples expressing the symbolism of the sidereal sign of Cancer.

Table 14.1 identifies the four 2150 year ages that span the 8,600 year continuum of this sidereal sign's different expressions of its symbolism. We will consider how love created new expressions of physical security, emotional nurturing, mental learning and spiritual blessings during successive 2150 year periods.

In the period between c. 9860-7710 BC when the last Ice Age was ending, the melting ice contributed to the eventual 120 metre rise in sea level and the flooding of thousands of miles of coastal lands across the world.[5] Earlier these changes were associated with the necessity of the communities forced to leave their homelands to find new ways to survive. The love of a *secure* home motivated people to search for a safe place to live and to learn the new skills demanded by the changing environments being established. This warmer and wetter climate released the fertility of vast tracts of formerly frozen lands and greater quantities of natural food became available. Eventually the growing population became too large to be supported by this supply of food. After c. 7710 BC, farming spread to manage the crops and livestock required to *nurture* growing populations. Earlier, these opportunities were considered in the context of the movement of

---

[3] New Larousse Encyclopaedia of Mythology, p.13.

[4] Bernadette Brady, Brady's *Book of Fixed Stars*, p. 257.

[5] Steven Mithen, *Prehistoric Europe*, p. 89.

surplus foods between communities as the first step towards the growth of trading cultures.

After c. 5560 BC, the spread of farming and the greater availability of food expanded opportunities for people to survive without traditional hunting or agricultural skills. The natural aptitudes able to be linked to a family's sense of identity, be they potter, carpenter, mason, smith or weaver may be mentally *learned* by successive generations. The emotional bonds within a family enhance the child's capacity to learn, and are illustrated by the way children naturally learn languages during infancy. Individuals learn what they are good at. In this case, learning becomes a process of the self-realisation of a particular natural talent. In extreme cases, a child may be identified as being gifted. This subject needs to be distinguished from mental education where formal teachings establish standards of behaviour and levels of skill people need in order to live and contribute to their culture.

In Mesopotamia the management of meat supplies, the irrigation for the cultivation of grain supplies and the availability of high status crafted copper products illustrate skills showing cultural diversification. The war-lords of this region became the leaders controlling the resources of separate City States.

The final phase of the continuum of Cancer began c. 3410 BC and the spiritual potential of this humanistic age indicates that evidence for the presence of *blessing* will be found. The first example comes from basic cultural differences between Mesopotamia and Egypt. It concerns the qualities of leadership that were developed and refined during the millennia of the humanistic Age of Leo, which preceded 3410 BC. The religious and secular leadership associated with the different City States of southern Mesopotamia appear to be different from Egyptian practices. It is possible that the subject of kingship was addressed with an agenda for a very different form of leadership. In Egypt, the idea was born of tribes being loyal to kings and kingdoms being created. For evidence that the royal objective was to bring spiritual *blessings* we refer to Prof. Stephen Quirke's insights from his book *Ancient Egyptian Religion.* The cult of kingship was central to the royal rule of ancient Egypt. The fundamental theme of this regal supremacy was for the nation to offer to heaven much of its produce and wealth. In exchange for these offerings, the king brought heaven's blessing to Egypt. [6]

The first Kings of Egypt and the early pharaohs were charged with the spiritual responsibility to act and heal creation. This necessity was believed because of humanity's fall from divine grace and the withdrawal of the sun god.[7] The work of the king was to heal all Wrong and cause Right (the religion of Matt) to exist throughout their kingdom.[8] The early Egyptian kings achieved divine status as sons of Ra and identified as being at one with the sun god as both the creator and all created substance.

The circumstances that led to the Upper and Lower Kingdoms of Egypt becoming the first nation of the world are not clearly described.   However, many themes have been identified that provide insights for us to consider. We want to see if the spiritual blessings identified above created *blessings* that manifested in mental, emotional and physical dimensions of Egyptian life.

From Table 13 we see that for the mental dimension we will be looking for expression of the symbolism of Gemini, for the emotional dimension that of Taurus and for the physical phase the symbolism of Aries. In the following interpretation of ideas about the creation of the Egyptian nation, spiritual, mental, emotional and physical themes are integrated into one scenario.

The historical evidence for describing the process that united the tribes of Egypt's Upper and Lower Kingdoms into the Egyptian nation is limited. Themes from this period suggest that the cult of the sun god arose from Egyptian beliefs of the power in heaven of the daylight Sun. The representative of this power on

---

[6] Stephen Quirke, *Ancient Egyptian Religion*, p. 81.

[7] Ibid p. 31.

[8] Ibid p. 32, 36, 38, 43.

Earth was described by the metaphor of the soaring falcon, the earthly ruler of the skies. Egyptian leaders used mental intelligence to introduce the myth that the falcon was the symbol for Horus, the divine responsibility of kings and later pharaohs to heal creation. The first Egyptian kings took this name as the supreme royal title.[9]

The imagery of the falcon flying above all the sacred tribal creatures illustrates a unifying national *blessing*, a deity that did not threaten tribal totems. The relationship each tribe had with its sacred animal was a central theme of an emotional dimension of Egyptian culture. Within the physical dimension, each tribe's totem was nationally protected within the over-arching belief system established by the divinity of royal authority.

In Chapter 5, we looked at the invention and use of writing in some detail. From this, we recognise how *education* played a central role when a written language was used for religious and secular matters. The actions taken for tribal lands to physically unite and create one nation, involved changing tribal beliefs. The emotional security this created has to be understood. The images used on ideograms and hieroglyphics illustrate the way Egyptian leaders used mental intelligence to create a written language from familiar aspects of life.

There were many other ways that emotional *blessings* were experienced during and after the unification of Egypt's Upper and Lower Kingdoms. New emotional values shaped the artistic expression of grave goods. These were originally provided as resources for use by the deceased in the afterlife. Around the time of the unification of Egypt, a new emphasis was placed on the value of the craftsmanship and rare materials buried with powerful personages.[10] During the previous age, the dating of distinct designs of pottery included goods of high quality, and is classified as Naqada I, II and III after their region of manufacture and is used to date many Nile sites when found in graves. When

---

[9]   Prof. Stephen Quirke, *Ancient Egyptian Religion*, p. 21.

[10]   *The Oxford History of Ancient Egypt*, The Naqada Period by Beatrix Midant-Reynes, p. 53.

the practice of high value goods began c. 3,500 BC, other styles of pottery in use at the time were found in the poorer region of the cosmopolitan Nile delta as well as upstream of Naqada at Nubia.

Naqada was located close to the mouth of the Wadi Hammamat, a tributary of the Nile. The savannah pastured valley of this Wadi led eastward towards quarries and mines. The resources of gold from this region provided the valuable raw materials for local leaders and the craftsmen of Naqada. Less than one hundred kilometres to the south, the community of Nekhen was also a centre for the gold industry. Here the supplies that came from the nearby Wadi Abbad were also mined from the Black Mountains found in the Eastern Desert. [11]

The change from the Naqada II to the Naqada III period, and the start of Dynasty 0 is commonly used to denote the time when King Narmer created the nation of Egypt, c. 3,200 BC.[12] In Lower Egypt at North Saqqara, the evidence of grave goods supports the idea that by 3000 BC the centre of national administration had been established at nearby Memphis. This location, strategically placed just before the Nile divides to form the delta region, was close to the former boundary between the Upper and Lower Kingdoms. The practical value of this site was also convenient for coastal trading. To the east of the delta at Maadian was the important centre of the copper industry.[13] The magnificent graves of the kings continued to be at their traditional location of Abydos in Upper Egypt.

Physical *blessings* in a national context must include the ways the potential of the river Nile were realised. Its resource as a national agricultural asset able to feed its population was only realised when the annual floods were managed for irrigation purposes. In addition, its value as Egypt's primary communications artery is difficult to overestimate. Another physical *blessing* was the building of the Egyptian Old Kingdom's

---

[11] *Genesis of the Pharaohs* by Toby Wilkinson, p. 124.
[12] *The Oxford History of Ancient Egypt*, The emergence of the Egyptian State, Kath. Bard, p. 65.
[13] Ibid

cities, temples and pyramids together with the capacity of its different regions to cooperate with their construction. The vast investments of resources required over many years to complete these buildings illustrate the value placed on the purposes they served. When the tribes of Egypt's Upper and Lower Kingdoms were united, the new 'White Wall' national capital city Memphis was built. The god Ptah was worshipped there as 'the Universal Demiurge', the 'Master Builder', 'who with his own hands fashioned the world', for protecting all artisans and artists.[14] His cult animal, the sacred Apis Bull was worshipped as 'the Renewal of Ptah's life'.[15] The Apis Bull's association with creativity extended through physical, emotional, mental and spiritual dimensions. Insights concerning this extent of the creative power Ptah symbolised come from his temple texts. These precise and subtle words describe the praise of Ptah as the creative power of the heart and tongue of Ra, the source of the Creative Word.[16]

The artistic expression of skills embodied in public buildings and works of art required the motivation to express the highest quality of work. The most famous contributor to public works was the sage and chief minister Imhotep, c. 2600 BC, deified by later generations as the son of Ptah for expressing the pinnacle of practical spiritual endeavour. It is understood he contributed to the design of the stunning Step Pyramid at Saqqara and possibly influenced that of the Great Pyramid of Giza. The *blessing* that came to Egypt via the early pharaohs stemmed from the *actions* taken by the culture they initiated and maintained.

This chapter's descriptions of the introduction of new physical, emotional, mental and spiritual dimensions of experience during the humanistic Age of Cancer are based on historical evidence. We will now test this technique over dozens of millennia to see if our model's precessional humanistic structure's words describe the much earlier creativity of human hearts and minds.

---

[14] New Larousse Encyclopaedia of Mythology, p. 34.
[15] Ibid, p. 43.
[16] Stephen Quirke, *Ancient Egyptian Religion*, p. 45.

14

The Fate of the Feminine

The earliest evidence of a cultural interest in the goddess comes from the discovery of artefacts known as the 'Venus' figurines dated between c. 27,000-22.000 BP.[1] A shaped and engraved bone plaque from south west France has been interpreted to be a possible lunar 'calendar' dated between c. 34,000-32,000 BP.[2] These examples of Upper Palaeolithic art were discovered across Europe in an arc stretching from the Pyrenees to southern Russia. These palm sized stone statuettes and ivory carvings illustrate various portable images of feminine fertility suggesting childbirth and mothering; they could be easily carried for personal use.

In the Foreword to the *Language of the Goddess* by Marija Gimbutas, Joseph Campbell writes that these simple feminine aids reveal how humanity's primordial life aspired to understand and live in harmony with Creation's archetypes. He describes how Gimbutas writes about the strong beliefs relating lunar and female cycles and their association with the continual regeneration of life after death traumas. This continuity, intrinsic to life's natural unity is symbolised by the numerous goddesses. All the different life forms that create the natural environment had goddess mythology to contribute to the cultural awareness of Earth's sacred mysteries.

The decline in goddess worship Campbell writes of begins when patriarchal cultures saw a linear route between birth and death in place of the feminine awareness of nature's cyclical continuums. With this change, cultures began to prioritise the

---

[1]   Paul Mellars, The Oxford Illustrated History of Prehistoric Europe, p. 72.
[2]   Ibid, p. 71.

growth of their power and later to improve their economic performance to provide greater opportunities to limit their vulnerability to powerful enemies and natural disasters such as floods, famines and diseases. To provide some idea as to the vast number of ways tried to achieve this objective during the early historic period, the book *The Golden Bough* describes the many subjects that attracted and guided such cultural changes. At the end of this abridged edition of this history of myth and religion, Sir James Frazer (1845 – 1941) includes a farewell. 'For the present we have journeyed far enough together, and it is time to part. Yet before we do so, we may well ask ourselves whether there is not some more general conclusion, some lesson, if possible, of hope and encouragement, to be drawn from the melancholy record of human error and folly which has engaged our attention in this book.'

Before the period studied by Frazer, the Goddess culture was yet to retreat to the forests and mountains. The 'Venus of Laussel' c. 19,000 BC bas-relief sculpture in a Dordogne cave is widely believed to represent the 'Great Earth Mother', a deified Upper Palaeolithic godhead.[3] The dating of this variety of art suggests the timing and stage of new and widespread beliefs concerning feminine cultural priorities. These dates indicate that many millennia are associated with the development of a feminine spiritual culture likely to be devoted to the worship of this Earth Mother deity. Her widespread veneration possibly fostered the gathering together of women in support of each other as mothers, and encouraged a matrifocal tribal life that emphasised the survival of infants. The purpose of this art has attracted interpretations that range from emphasising feminine virtues to their use as tokens exchanged across a network of Ice Age communities.

The introduction of this culture as the last Ice Age was reaching its coldest period may be associated with new survival challenges during the millennia before and after the c.18,000 BC

---

[3] Norman Davis, *Europe A history*, p. 72.

glacial maximum. Another 12,000 years passed before the warmer and wetter climates settled into their annual rhythms. The numerous environmental upheavals during this warming period were driven by the greater presence of floodwater and rain. The surges of floodwater created by the ice melts raised the sea level of the world's oceans by over 120 metres. Forests expanded northwards bringing flora and fauna to live on the former frozen tundra.

The cultural traditions that were established during the last Ice Age had to adapt to the changing conditions that came with the warming climate. Eventually the traditions reflected the new skill demanded by the warmer environment. One successful culture moved from its traditional homeland when it followed the reindeer herds as their patterns of migration moved northwards from southern France to Scandinavia. Organised communities that supported each other may have been more successful than small independent family groups when faced with the challenge to change.

The artwork describing goddess cultures during the Holocene, which followed the Upper Palaeolithic after 10,000 BC, expanded as its focus attended to many themes relating to the Earth Mother Goddess imagery. To explore the extent of the cultural interest in numerous goddesses we again refer to the comprehensive works of Gimbutas, which include the titles *The Goddesses and Gods of Old Europe* and *The Language of the Goddess*. While the culture that identified with the feminine subjects of fertility and motherhood continued throughout and after the Neolithic, the Earth Mother Goddess came to be associated with the fertility of the natural world and regeneration of all life on Earth. To understand her ways included the veneration of the laws of nature and the mysteries of life and death. 'She was the single source of all life who took her energy from the springs and wells, from the Sun, Moon and moist earth'.[4]

---

[4] Marija Gimbutas, *The Language of the Goddess*, Introduction xix.

Marija Gimbutas describes how the Neolithic Revolution's introduction of agriculture to south Eastern Europe included artistic developments that were more advanced than those from surrounding regions were. The heart of this 'Civilization of Old Europe' extended from the eastern coast of the Adriatic through the Danube's catchment area to the Black Sea's western shore and from Czechoslovakia in the north to Crete and Sicily in the south. The Goddess religion of this region described by Marija Gimbutas is interpreted as part of 'cohesive and persistent ideological system' spreading across the Near East, around the Mediterranean, from central to northern and Western Europe. Many of the goddesses are identified by common symbols like the Chevron, the V or Zigzag patterns. Other sacred symbols and images of goddesses, their birds, animals and mysterious creatures brought meaning to normal life in Old Europe. The richness of gold artefacts at Varna funerals in eastern Bulgaria was greater than any found in the near East c. 4300 BC. [5]

After this date the matrilineal goddess culture of Old Europe began to be undermined by the patrilineal warrior gods worshipped by Russian settlers from the Pontic Caspian. The goddesses that were lost were replaced in later millennia when Persian mythology introduced the names of Ishtar, Inanna and the underworld's Ereshkigal. Later Greek mythology contributed the goddesses of Olympus, Hera, Athene, Artemis, Hestia, Aphrodite and Demeter. The goddesses Persephone and Hecate came with insights about the hidden powers from life's depths. The roles of these goddesses in the patriarchal cultures of Persia and Greece often supported the myths of cultural heroes such as Gilgamesh and Hercules. The mythic dramas of the goddesses were part of the Athenian tragedies that address the poignant dramas of human experience the modern mind knows well.

In chapter 5, we saw that this loss of the goddess was

---

[5]   David W. Anthony, The Horse, the Wheel and Language, p. 225.

forecast in ancient Egyptian religion even when the purpose of the cult of the Sun god, to preserve the universe and heal creation, was protected by many different goddesses. Is it possible Egyptian sages were aware of what was causing the demise of goddess worship? Before assessing their capabilities to understand this we will consider some insights about the departure of the goddess provided by a brain surgeon intrigued by the many lost goddesses of ancient Europe.

It is common knowledge that the left hemisphere of the brain controls the right side of the body while the right hemisphere controls the body's left side. It is less well known that the brain's left hemisphere greatly influences speech and follows the laws of grammar and logic while the right hemisphere's capacities include recognising faces, intuition and the language of feelings. This difference has led to each brain's hemisphere being identified as either masculine or feminine. The brain's masculine or left hemisphere remembers the rules and structural details of languages. The brain's right hemisphere, its feminine side, is used when we look at pictures.

Understandings that are more recent have developed from the idea known as the lateralization of brain functions. These indicate that the above cerebral polarity is far more subtle. However, we do not know if the nuances associated with the modern mind can be accurately applied to ancient times when life was less complex.

Leonard Shlain, writing about brain functions from his perspective as a brain surgeon considers the implications of the left-brain right brain polarity. In his book *The Alphabet versus the Goddess* Shlain argues how writing has changed the way the brain works. His observations draw on 'brain anatomy and function, anthropology, history and religion'. These describe how literacy exercised the brains left hemisphere with a reliance on memory and logic. These masculine qualities excluded the use of the feminine right hemisphere's intuitive, visual, holistic capabilities. Their use initiated the decline in goddess worship together with the rejection of imagery. As patriarchal powers increased, the goddess virtually disappeared together with the status of women.[6]

Leonard Shlain argues that the cultural use of writing has resulted in actual physical changes in the brain over time. A psychological polarity developed between the word and image; emphasising the letter of the word at the expense of viewing an image changed the gender balance within literate cultures. The use of the logic needed to read and write requires the brain's left hemisphere to become more active by using the learning that has taken place on all manner of subjects. The work involved in all this doing takes place at the expense of the right hemisphere's capacity to see a complete picture in an instant. This capacity of being is expressed when intuition knows the truth or suddenly knows the right answer. An acceptance of 'what is' replaces the personal focus on 'what may be'.

The gender polarities within many different cultures studied by Shlain illustrate how a dependence on the written word's power established masculine priorities at the expense of feminine values. The laws of grammar, which all writing has to adhere to, are learnt and establish the abstract rules the left-brain uses. These reinforce masculine principles and are the antithesis of the spontaneity and intuition of feminine wisdom.

Shlain observes some implications, which arise from the use of Egyptian hieroglyphs when compared to Mesopotamian cuneiform script. Hieroglyphs were read as images, whereas cuneiform script developed into complex abstract symbols that had to be learnt. This fostered a Mesopotamian elite social class of scribes with highly developed left-brain skills. The dominant theme of the Mesopotamian 2350 BC Hammurabi law code further established a culture of patriarchy, logic and truth. An early milestone was created where the written law increasingly displaced cultures that valued the subjectivity of intuitive truth, the wisdom of the Earth Mother Goddess. To understand her ways included the veneration of the laws of nature and the

---

[6]   Leonard Shlain, The Alphabet versus the Goddess, cover.

mysteries of life and death.

The growth of the left-brain's skills accelerated the decline of goddess worship. The cultural impact of this prioritised the scientific advances and the intellectual knowledge needed to exploit the Earth's natural resources. During recent millennia, education has powered the industrial destruction of our planet's continuing capacity to support human life. While ancient Egyptian culture knew nothing of these details it seems that their cosmology confirmed this danger.

Clues about foreseen problems are mentioned in the Pyramid Texts where reference is made to the immortal ones, the spirits of the imperishable circumpolar northern stars that never set. The details of this observed stellar permanence are a feature of the centre of rotation for the sky, the celestial North Pole, the point in space located 30° 03' above the northern Egyptian horizon. The observed rotation of these stars is a result of the Earth spinning on its axis. This centre of rotation of the sky is like a bearing in which the axis of the Earth spins. Within a circle formed by the radius created by the distance from the horizon to this northern celestial central point, the stars never set. They are free from the daily cycle of rising (birth) and setting (death). As such, the Egyptians associated these imperishable stars with eternity and immortality. This was a vital dimension of their cosmology since much of their culture pivoted on the quality of the afterlife of their pharaohs. For some insight on the funerary drawings and religious texts about the afterlife destiny of kings and nobles, we note the words of the American Egyptologist James H. Brested. The northern circumpolar stars of the Plough in Ursa Major, of Ursa Minor (the Little Bear) and Draconis (the Dragon) were believed by ancient Egyptians to be the hosts for dead kings.[7]

When the Upper and Lower Kingdoms united to create the Egyptian nation c. 3,100 BC, the precessional movement of the

---

[7]   Robert Bauval, The Egypt Code, p. 240.

circumpolar stars was bringing the alpha star of the sidereal sign of Draconis, the Pole star Thuban, into exact alignment with the celestial North Pole. Astronomical observations have determined that this position moves very slowly across the sky to trace a complete circle in the heavens every 25,800 years. This second way to identify precessional movement is shown by the behaviour of a child's toy spinning top as it slows down. Just before it topples over and stops, its vertical axis begins to gyrate. This axis traces a circle in the air above the top. We use this model (see page 82) to illustrate how the axis of the spinning Earth is tilted at 23° 28' from the vertical and points to a place in the heavens called the celestial North Pole. This gyration of the Earth's axis very slowly traces a circle amidst the stars. This circle shows the path followed by the celestial North Pole. When this celestial pole is close to a star, this point of celestial light becomes the northern Pole Star. The symbolism of this star and its sidereal sign gives us some idea about the processes taking place on Earth over the millennia.

The demise of the goddess is a theme Bernadette Brady writes about in her description of the sidereal sign of Draco.[8] For many millennia, this feminine centre of the heavens has been symbolised as dragon, serpent or snake by different cultures as the feminine guard and protector of creation. The fundamental relationship between them and the Earth Mother or the divine feminine is explained by the positions of the celestial poles above the Earth's northern and southern hemispheres. They ensure the axis on which the Earth spins tilts towards the Sun in summer and away from the Sun in winter, see Fig 8.1. This cycle of changing exposure to sunlight provides a basic rhythm that supports the cycle of the seasons.

We can learn of the danger that threatened this region of the heavens from the symbolism of the later Babylonian astronomer/priests who invented their own zodiac by using the

---

[8]   Bernadette Brady, Brady's Book of Fixed Stars, p.133.

annual path of the Sun through its twelve sidereal signs of stars. This belt of stars around the heavenly vault was observed when addressing the terrestrial dramas below, as symbolised by the movements of the planets. This zodiac was rich with omens as the various planets were observed to rise into view, ascend into prominence, and then disappear from view. These wanderers never approach the revered point of the sacred centre of feminine creativity. The Babylonian mythologies describing the meanings of the northern stars are ripe with omens of danger. Mesopotamian star watchers saw these northern circumpolar stars as the dragon guard, nightly patrolling around this sacred region of the heavens with the responsibility for protecting the divine feminine.

To understand the dangers faced by the sacred centre of the heavens we will study the centuries after 3000 BC in more detail, when Thuban was the Pole Star. At this time, the two Kingdoms of Egypt had just been united and the foundations were being laid for the Old Kingdom with its famous pyramid culture. Were the Egyptian seers aware of any problem or threat to the centre of the heavens? The evidence gleaned from the Great Pyramid by many specialists has been interpreted to indicate that they knew a great deal more than this. The discoveries made by academic teams in order to understand more about the Great Pyramid identified all manner of stellar alignments with its passages and ventilation shafts. A feature, which continually arises in these studies, is that its architect had knowledge of precession and designed this pyramid for use with the celestial North Pole and the alpha star of the sidereal sign of Draco, Thuban. This star introduces to us the idea that its symbolism may contribute to our understanding of dangers to the Eternal Feminine that were emerging when the celestial North Pole was close to Thuban. In chapter 8, Fig 8.2 illustrates the path of the celestial North Pole and some stars which were or will be the northern Pole Star.

The details of the Great Pyramid's external and internal measurements are considered by Davidson and Aldersmith in their book *The Great Pyramid – Its Divine Message,* published in 1925. A basic observation they make is that the pyramid's designer had intended that it's 'symbolism was to be interpreted in an age already in possession of the astronomical knowledge embodied in the symbolism projected.' Moreover, 'We are compelled, then, to come to the conclusion that the Pyramid's external features were designed to attract and direct attention to a further message of greater importance'.[9] This pyramid's pattern of internal passageways has been interpreted to describe humanity's journey through history *after a date* derived from the architecture of the Great Pyramid. Details of this history begin at the Great Pyramid's entrance in its northern face. This descending passage slopes downwards at an angle of 26° 18' 9.7"and is almost four degrees out of alignment with the northern celestial pole's elevation of 30° 03' north. However, in 2141 BC precessional movement brought the Pole Star into exact alignment with the Great Pyramid's descending passage.[10] Thuban's light penetrating the depths of the pyramid's subterranean chamber, symbolised testing and a downward journey for humanity.

At the time of this alignment, the catastrophe that overtook Egypt centred on the failure of the Nile flood for several years. This river was the traditional symbol for the goddess Isis, which could not provide the harvest that supported life in Egypt. The extent of this disaster can be assessed by the fact that the Faiyum, a lake 65 metres deep, dried up.

Recent discoveries reveal that this global cooling was caused by lack of rain and was part of a climatic pattern that existed between 2200-1890 BC. It extended from Iceland to Nigeria

---

[9] Davidson and Aldersmith, The Great Pyramid Its Divine Message, p. 139.
[10] Peter Lemesurier, *The Great Pyramid Decoded* , p. 48.

and included findings from Tibet, Italy and England.[11]

From this consideration of cultural changes taking place in the context provided by the precessional cycle we note that the conscious appreciation of feminine goddess imagery began c. 27,000 BC. This date indicates that human consciousness has had a complete precessional cycle to develop an appreciation of many qualities of the goddess before masculine spirituality and monotheism became the dominant model. The link between the development and refinement of feminine spirituality and some 26,000 years later the start of the development of cultures that value masculine spirituality illustrate an alignment between successive precessional cycles. This synchronicity between these successive precessional cycles suggests other themes may be related through cyclical correlations. Evidence of this is provided by the dates between the Upper Palaeolithic Revolution, c. 38,000 BC and the Neolithic Revolution, c. 12,000 BC. In the next chapter, we will first assess how far back in time the evidence of cultural change can be traced. Then the near 26,000 year period between these two revolutions will be explored. We will see how accurately a complete cycle of precessional symbolism correlates with the known evidence of cultural change that took place during this period.

[11] Professor Fekri Hassan, Ancient Apocalypse: The Fall of the Egyptian Old Kingdom; BBC, June 2001

15

Stone Age Enquiries

Archaeologists and palaeontologists have long discussed the relative significance of a wealth of artefacts and fossilised bones from sites across the world. Great care has been needed to piece together the sparse evidence to form an understanding of many basic aspects of how our ancestors lived. Evidence from the last six million years reveals how the Hominid family of primates has evolved into several different genii and sub species, classified by their different features. Evidence of the use of hand axes alongside fossilised remains led to the identification of *Homo habilis,* alive about 2 million years ago. From this first genus (*Homo),* other species developed with larger brains: *H. erectus* c.1.8 million years ago has been found in fossil form. After another c. 1.3 millennia, it is believed that these two developed into two new species *H. heidelbergensis* and *H. sapiens* and that these in turn led to *H. neanderthalensis* and *H. sapiens sapiens* respectively. This last species has our skeletal structure and originated around 100,000 years ago in Africa, and subsequently spread through Asia, Europe and the Americas.

How far back can we go for evidence of changes in Stone Age consciousness? This question is answered in the book *The Neanderthal Legacy,* written by Prof. Sir Paul Mellars from Cambridge University and published in 1995. As a distinguished authority on the Middle and Upper Palaeolithic Stone Age periods, Mellars' writing describes how our ancestors began to migrate from the Near East into Europe c. 40,000 BC and gradually displaced the Neanderthals. The artefacts made during this time illustrate the first evidence of the human imagination at work. This development is used identify the start of the Upper Palaeolithic Revolution. His writing in the book *The Oxford Illustrated History of Prehistoric Europe* edited by Barry Cunliffe, further describes this revolution. These books provide two sources used in the following chapters.

A vital theme of this evidence is the method used for its dating. The older the artefacts the greater the risk that the radiocarbon BP (Before Present) date is many thousands of years away from its calendar date. This happens because radiocarbon dates are calculated by measuring the carbon-14 radioisotope contained in an artefact. After death, the physical remains gradually lose this radioisotope. However, there is the risk of the minute quantities of carbon remaining being supplemented by contamination absorbed from a normal isotope rich environment. The ability to calibrate accurately carbon-14 dates into calendar years depends on the amount of C14 in the atmosphere being known when the radiocarbon isotope began to decay. Many different calibration curves have been created to determine the calendar year age of various materials from different locations. Ways to approach these problems were agreed at the 19th International Radiocarbon Conference at Oxford in 2006. The evidence, with BP dates considered below, is converted to BC dates by using the IntCal09 calibration curve.

The need for cautious attention when considering radiocarbon dates is given in the following caveat from Clive Gamble in *The Oxford History of Prehistoric Europe*: 'It has become clear that radio carbon dates before 10,000 years ago may in fact be significant underestimates of *time* ages-probably by around 1,000 years at c. 10,000 BP, and perhaps by as much as 3,000 years at 30,000 BP. [1]

Between c. 45,000 BP and c. 33,000 BP, the newcomers gradually spread from the lands of the eastern Mediterranean

---

[1] Clive Gamble, The Oxford Illustrated History of Prehistoric Europe, page 41.

coastal region northwestwards into Central Europe, northern
Spain and then south-western France to start a remarkable
period of progress.[2] Different sites north and south of the Alps
and east and west of the Pyrenees have revealed that this
species developed their own Stone Age technology during this
period. At the same time, the European Neanderthal population
declined.   From what was probably their last European
occupied site in central France evidence of advanced tools has
been found dated c. 35,000 BP.[3] These artefacts have been
interpreted to indicate that before they became extinct Europe's
archaic residents learned new skills directly or indirectly from
our ancestors. Other Neanderthal sites also exhibit this cross-
cultural influence.

From sites in southwestern France, occupied after c. 40,000
BP by anatomically modern humans, artefacts were discovered
that revealed new expressions of creativity that warrant being
regarded by many as having been revolutionary.[4] The new
tools in question are recognised as having several distinct
features. In addition to a greater amount of work invested in
finishing their flint shapes, there was a wider variety of specific
designs. These changes reveal the difference between the skill
to make sharp tools and those needed to make tools to an
imagined shape.

The limitations of using brittle flint as the raw material for
the work imagined, were widely lifted as bone, antlers and
ivory were carved. The 34,000 BP - 33,000 BP dates of French
examples of this work from the 'Chatelperronian' levels convert
by using IntCal09 to c. 37,000 BC - 35,500 BC.[5] The use of
imagination to create early forms of art was to become a feature
of growing significance: the direction of our evolution as

[2]   Paul Mellars, *The Neanderthal Legacy*, page 405-10.
[3]   The Oxford History of Prehistoric Europe, ed. Barry Cunliffe. The Palaeolithic Revolution by
Paul Mellars, page 57.
[4]   Ibid, page 42.
[5]   Ibid, page 49.

revealed by modifications in skeletal structures was now being assessed by the changes introduced by artistic dimensions of consciousness. Many insights into this phenomenon have been obtained by anthropologists' interpretations of the way in which the practice of new skills and the artefacts crafted reflect the mind's early expression of an imagination. The widespread discovery of different types of artefacts coming into common use during the Upper Palaeolithic in different geographical regions of Europe provides pivotal evidence of the progress we seek to consider.

To assess if the cultural changes associated with the Upper Palaeolithic Revolution correlate with the symbolism of the precessional cycle we will extend the period earlier aligned with this cycle into the archaeological record of this prehistoric period. In Chapter 12, we correlated the historical record concerning monotheism with the date 1260 BC as depicted in Egyptian history by Akhenaten and Moses with the cycle of precessional symbolism. When we go backwards one complete precessional cycle of 25,800 years, the date of 27,060 BC is aligned with this cusp. It correlates with the introduction of the 'Venus' figurines we related to the spread of the Earth Mother Goddess culture. This same point in the precessional zodiac aligns with the start of the humanistic Age of Gemini within the Upper Palaeolithic.

This correlation provides the first prehistoric validation of the 1260 BC alignment with precessional symbolism. To check the alignment of the Upper Palaeolithic archaeological record with precessional symbolism we move in 2150 year periods from 1260 BC to apportion the precessional symbolism for all the humanistic ages of the Upper Palaeolithic.

The richest collection of Upper Palaeolithic artefacts originating after c. 40,000 BC came from many Upper Palaeolithic settlements established in the river valleys of the Garonne in southwestern France. Evidence from this region suggests that an abundance of horse, bison, migrating reindeer and other hunted animals provided a rich source of food whereby the human population increased to perhaps five or even ten-fold. [6] This region of France has provided the largest body of evidence showing how anatomically modern man settled in western Europe, leaving evidence of how his daily activities differed from those of the Neanderthals that had earlier occupied the same region, often the same sites.

The starting dates of humanistic ages aligned within the Upper Palaeolithic are shown on Tables 15.1 and 15.2 below.

Table 15.1

Overview of Early Upper Palaeolithic Humanistic Ages starting dates BC							
Sidereal signs and Themes	Sagittarius 39,960	Scorpio 37,810	Libra 35,660	Virgo 33,510	Leo 31,360	Cancer 29,210	Gemini 27,060
Sagittarius expansion	*belief*						
Scorpio power	*transforming*	*rebirth*					
Libra relating	*inter-dependence*	*agreement*	*goodwill*				
Virgo purpose	*serve*	*care*	*discriminate*	*purity*			
Leo radiance		*identity*	*confidence*	*centre*	*self-awareness*		
Cancer love			*security*	*nurturing*	*learning*	*blessing*	
Gemini connecting				*movement*	*intuition*	*educate*	*inspiration*
Taurus fertility					*resources*	*values*	*answers*
Aries strength						*action*	*courage*
Pisces, non-separation							*acceptance*

[6] The Oxford Illustrated History of Prehistoric Europe, Paul Mellars page 64.

The humanistic ages of the Upper Palaeolithic after those included in Table 15.1 are extended below on Table 15.2 into show the millennia when the Neolithic Revolution began.

The millennia spanned by these two tables are of interest to us since they illustrate another example of cyclical synchronicity. To show this we refer to evidence considered in Chapter 3. The archaeological studies from Abu Hureya in Syria concerning domesticated grain included in *Foraging and Farming* by G. C. Hillman et al. provide the required information.

Table 15.2

Overview of Late Upper Palaeolithic Humanistic Ages starting dates BC							
Sidereal signs and Themes	Taurus 24,910	Aries 22,760	Pisces 20,610	Aquarius 18,460	Capricorn 16,310	Sagittarius 14,160	Scorpio 12,010
Taurus fertility	*aspiration*						
Aries strength	*initiating*	*will*					
Pisces non-separation	*compassion*	*contemplate*	*unity*				
Aquarius evolve	*group*	*friendship*	*intelligence*	*ideal*			
Capricorn control		*establishing*	*authorising*	*knowledge*	*sacrosanct*		
Sagittarius expansion			*opportunity*	*honesty*	*truth*	*belief*	
Scorpio power				*releasing*	*sharing*	*transform*	*rebirth*
Libra relating					*evaluation*	*inter-dependence*	*agreement*
Virgo purpose						*serve*	*care*
Leo radiance							*identity*

The oldest samples of this site's rye grain crops are dated c. 11,140 BP, which converts to the calendar date of 11,050 BC. [7]

From Table 15.2 we see this evidence was in the humanistic Age of Scorpio, which commenced in 12,010 BC. When we move back 25,800 years to the same position in previous precessional cycle see Table 15.1, we arrive at the date of 37,810 BC, (c. 40,250 BP).[8]

This date is very close to the estimate for the start of the Upper Palaeolithic Revolution given by Paul Mellars from the ivory, bone and antler artefacts dated c. 34,000-33,000 BP.[9] These calibrate to c. 37,000-35,500 BC using the IntCal09 calibration curve.

The evidence of many of the cultural changes during the transition from the Middle to the Upper Palaeolithic is shown in Table 15.3 below. These archaeological artefacts and events have provided valuable insight about the cultural changes of the period. It lists some of the radiocarbon carbon-14 dates used to identify the start of the Upper Palaeolithic. The selection of bones, antlers, ivory and teeth used for the manufacture of different types of artefacts illustrates the new use of familiar animal parts formerly discarded. Although not as hard as flint, they have the advantage of not being brittle and allow the shapes imagined by craftsmen to be accurately replicated in durable materials for which Mellars uses the term 'prehistoric plastic'. The evidence of many advances in the manufacture of stone tools are not included in Table 15.3 since these developments represent refinements of the earlier Middle Palaeolithic flint work practices.

When Mellars assesses some aspects of the early Upper Palaeolithic progress to argue that a revolution did take place,

---

[7] IntCal09 Supplemental Data, P J Reimer, p 1120.
[8] Ibid, p 1135.
[9] The Oxford Illustrated History of Prehistoric Europe, Paul Mellars page 49.

the artefacts described each have two BP radiocarbon dates to suggest the possible probability of their ages as shown in Table 15.3 below. There is only limited evidence from Bulgaria for a series of progressive steps, but this feature is of interest to us as it may be an example of original progress being made in isolation. It may have been from this place that progress spread to become part of a wider culture. A pattern in the way in which progress is widely adopted some millennia later beyond its region of origin, was described earlier. We observed that it took about 2,000 years from the time when Aristotle explored the new frontier for the development of the sciences to the wider cultural introduction of the education that led to the Scientific Revolution.

Table 15.3
Estimated calibrated dates (BP date to BC)
of artefacts from the Early Upper Palaeolithic period.

Early Stages of Upper Palaeolithic	Carbon-14 BP date estimates	Carbon-14 BP mid date estimates	IntCal09 Calibrated BC dates
The creation of animal-tooth pendants, Bacho Kiro, Bulgaria.	c. 40,000 BP	c. 40,000 BP	c. 42,000 BC
The creation of animal-tooth pendants, Central France.	35 - 33,000 BP	34,000 BP	c. 37,000 BC
Groups specialising in hunting specific animal herd species.	35 - 32,000 BP	33,500 BP	c. 36,500 BC
Settlements of hut structures.	34 - 33,000 BP	33,500 BP	c. 36,500 BC
Design of bone, antler and ivory artefacts.	34 - 33,000 BP	33,500 BP	c. 36,500 BC
The use of ivory beads.	34 - 32,000 BP	33,000 BP	c. 35,600 BC
Possible lunar calendar engraved on bone.	34 - 32,000 BP	33,000 BP	c. 35,600 BC
Prolific manufacture of personal ornaments.	34 - 30,000 BP	32,000 BP	c. 34,600 BC
Lion headed man and highly stylised carved symbols.	34 - 30,000 BP	32,000 BP	c. 34,600 BC
Chauvet Cave paintings of rhinos, lions, bears & mammoths.	32 - 30,000 BP	31,000 BP	c. 34,500 BC
Venus figurines.	27- 22,000 BP	24,500 BP	c. 27,500 BC

All listed BP dates on table 15.2 are quoted from The Oxford Illustrated History of Prehistoric Europe, edited by Barry Cunliffe; Chapter 2, The Upper Palaeolithic Revolution by Paul Mellars. A mid date column is used to give a clearer estimate of the BP *order* of changes.

The cultural changes Mellars identifies as evidence of the Upper Palaeolithic Revolution includes the main cultural features briefly listed below.

1. The existence of a great number of cooperating individuals.[10]

2. Clearer organisation and design of living accommodation around a central hearth, better to serve the larger groups of inhabitants.[11]

3. The greater use of high-grade flint from more distant sources for particular tools, reveal the practice of discrimination.[12]   Stone-tool workshops were established adjacent to quarries and extraction sites indicating one practical expression of cooperation.[13] Bones, teeth, antlers and ivory were used to make products that cannot be made of stone.[14]   The later evidence of many more tools made to different designs illustrates the *beliefs* of toolmakers in their skill to create the artefacts they imagined.[15]

4. The discrimination to hunt particular species of game from those available.[16]

As the Upper Palaeolithic commenced a new European culture developed expressing a more sophisticated consciousness.

When we develop an overview of the archaeology of the period spanned by Table 15.1, we can estimate a date for the start of the Upper Palaeolithic Revolution by linking it to the start of the Neolithic Revolution, which began one cycle of precession later. On this basis, the humanistic Age of Scorpio provides the symbolism for the start of each revolution. Using

[10]   Paul Mellars, *The Neanderthal Legacy*, p. 378.
[11]   Ibid, p. 313.
[12]   Ibid, p. 266.
[13]   Ibid, p. 386.
[14]   Ibid, p. 371.
[15]   Paul Mellars, The Oxford Illustrated History of Prehistoric Europe, p. 45 - 53,
[16]   Paul Mellars, *The Neanderthal Legacy*, p. 231.

the date of 12,010 BC for the start of the Neolithic Revolution, the start of the Upper Palaeolithic period began 25,800 years earlier and provides us with the date of 37,810 BC. This use of the precessional cycle provides a natural reference date for us to use. The spiritual, mental, emotional and physical dimensions of each age have their keyword as determined in Chapter 10. To aid our consideration of the early ages of the Upper Palaeolithic an extract from Table 15.1 is given below as Table 15.4.

We now want to see if the artefacts described in Table 15.3 illustrate a new expression of creativity. Does it correlate with the precessional symbolism of the respective humanistic ages aligned with their date of manufacture, as shown below in Table 15.4? To assess this, we will use the likely intended purpose of each surviving artefact to describe the imagination of its maker.

We know that radiocarbon dates shown below are age estimates of varying degrees of accuracy; however, it is likely that the order of their ages is correct. This suggests the likelihood that pendants made from drilled animal teeth are some of the earliest evidence of the Upper Palaeolithic imagination. The keywords for the humanistic Ages of Sagittarius and Scorpio provide themes concerning the practical and symbolic use of the teeth of carnivores.

The drilled teeth artefacts cannot be directly related to survival needs, and show that time and effort were available to invest in them for some personal use: possibly adornment. It is suggested that their use emerged out of recognising the purpose served by the powerful teeth of carnivorous animals in comparison to human hunting weapons. A *belief* (Sagittarius) that teeth *served* (Virgo) hunting successes created the belief that teeth worn by a hunter increased their power to kill animals. These conditions allow the *rebirth* of the hunter's belief in their own *identity.*

Tooth necklaces and pendants could have had any number of uses: identifying members of a hunting group or specialist lone hunter, for example. Even for stone toolmakers to wear in the belief that they somehow conferred the power to shape stones as imagined. Irrespective of these possibilities, it is realistic for us to surmise that the use of the tooth pendant represented some belief directly related to the known hunting power of the original animal. This use of personal ornaments possibly enabled the wearers to experience some form of *rebirth* in their own eyes, or the eyes of others, as they displayed these symbols of power.

Table 15.4.

Humanistic developments during the start of the Upper Palaeolithic period with overlapping 2150 year phases.

Sidereal signs of the Zodiac	Sagittarius 39,960	Scorpio 37,810	Libra 35,660	Virgo 33,510	Leo 31,360
Qualities of Sagittarius	**spiritual** *belief*				
Qualities of Scorpio	mental *transforming*	**spiritual** *rebirth*			
Qualities of Libra	emotional *inter-dependence*	mental *agreement*	**spiritual** *goodwill*		
Qualities of Virgo	**physical** *serve*	emotional *care*	mental *discrimination*	**spiritual** *purity*	
Qualities of Leo		**physical** *identity*	emotional *confidence*	mental *centre*	**spiritual** *self-awareness*
Qualities of Cancer			**physical** *security*	emotional *nurturing*	mental *learning*
Qualities of Gemini				**physical** *movement*	emotional *intuition*
Qualities of Taurus					**Physical** *resources*

Where keywords are shown in the text, they appear in an italic font.

The large rise in the population of south west France together with the improvements in living accommodation identified by Mellars appear to correlate well with the spiritual *goodwill* of the Age of Libra and the emphasis on physical *security* during this period. Another theme Mellars mentions as being evident during this period of transition concerns the selection of specific animals that were hunted. This provides evidence of *discrimination* being exercised by hunters.

When looking at the humanistic Age of Virgo we will assess how the keywords that relate to the period after 33,510 BC introduce new themes. The ivory carving of a person with the head of a lion, found in a cave at Hohlentsein-Stadal has aroused much speculation about its ancient use. One idea is that it was used to enhance the hunter's belief in their ability to *move* like a lion to kill game. For such success, the hunter's mind has to be *purified* through the some sort of ritual or meditative practice where the mind is *centred* on the intended objective. The success of this expression of progress can be appreciated to depend on the hunter's ability to *nurture* his own abilities. In this process, we have the idea that hunters also provide for their communities with physically *security* (Cancer) from dangerous animals.

When considering the spiritual qualities aligned with the Age of Leo c. 31,360 BC and its associated *self-awareness,* we expand our scope for speculation by associating this statuette with a shaman/teacher/priest who may *intuitively* commune with lions, as he teaches his pupils to *learn* to become better hunters. The possibility of the powerful imagery of this statuette is aided by the presence of the other animal carvings found with it. These clues suggest its possible use in rituals. [17] It is possible a similar purpose was served by the Chauvet cave paintings of the famous lions in France c. 31,000 BP.

The earliest Chauvet cave art features dangerous lions, rhinoceros, mammoths and bears that were painted around the time the lion-headed man was carved. The later paintings of

---

[17] *The Mind in the Cave* by David Lewis-Williams, p. 202.

horses, bison, aurochs and deer depict the game animals on which people were dependent. This range of species provides a rich agenda for contemplation. If the motivation behind these drawings was the fertility of the natural environment, its capacity to provide the animal *resources* for hunting was clearly a subject not taken for granted by this culture.

Caring for the natural environment is an agenda attracting increasing attention today. It is widely recognised how the industrial exploitation of global resources continues to threaten the prospects of our own and future generations.

To appreciate the context of our own humanistic Age of Taurus, relative to the other ages of the Holocene, which began 10,000 BC, Table 15.4 is included overleaf. It identifies the physical, emotional, mental and spiritual dimensions of experience aligned by the precessional cycle. It illustrates the repetition of keywords used on Table 15.2 when this same age name was aligned to the period 24,910-22,760 BC. During these millennia, the last Ice Age was approaching its glacial maximum.

In this chapter we have considered a range of scenarios that illustrate how the artistic crafting of artefacts were part of the cultural changes that developed during the period c. 38,000 to c. 29,000 BC. This progress during the early part of the Upper Palaeolithic correlates with the zodiac symbolism of the precessional cycle aligned with this period. The next objective is to see how the continuum of progress correlated with the aligned precessional symbolism extends into our own times

.

Table 15.5

Overview of Humanistic Ages of the Holocene Period.
Estimated dates for the introduction of new qualities of consciousness.

Sidereal signs of the Zodiac	Holocene Humanistic Ages					
	Libra 9860 BC	Virgo 7710 BC	Leo 5560 BC	Cancer 3410 BC	Gemini 1260 BC	Taurus AD 890
Virgo purpose	mental discrimination	spiritual purify				
Leo radiance	emotional confidence	mental centre	spiritual self-aware ness			
Cancer love	physical security	motional nurturing	mental learning	spiritual blessings		
Gemini Connections		physical movement	emotional intuition	mental education	spiritual inspiration	
Taurus fertility			physical resources	emotional values	mental answers	spiritual aspirations
Aries strength				physical action	emotional courage	mental initiative
Pisces nonseparation					physical acceptance	emotional compassion
Aquarius reform						physical group

16

The Rise of Ice Age Cultures

We have traced the early evidence of Stone Age man exercising his imagination during the millennia between 38,000 and 30,000 BC; the early part of the Upper Palaeolithic. We will now explore the millennia that followed and led to the coldest period of the last Ice Age. The period between c. 30,000 and c. 20,000 BC provides much evidence of cultural developments in Europe benefitting from the use of the imagination.

Table 16 below shows the continuum of humanistic ages and the qualities of consciousness our model indicates were being expressed as the last Ice Age approached its coldest time.

Table 16

Humanistic qualities during the Middle of the Upper Palaeolithic

Sidereal sign of the Zodiac	Starting dates BC and qualities of Humanistic Ages within the Upper Palaeolithic					
	Leo 31,360	Cancer 29,210	Gemini 27,060	Taurus 24,910	Aries 22,760	Pisces 20,610
Leo	self-awareness					
Cancer	learning	blessing				
Gemini	intuition	education	Inspiration			
Taurus	resources	value	answer	aspiration		
Aries		action	courage	Initiating	will	
Pisces			acceptance	compassion	contemplate	transcend
Aquarius				group	friendship	intelligence
Capricorn					establish	authority
Sagittarius						opportunity

The keywords describing the humanistic qualities aligned with the ages identified in Table 16 come from the table of keywords given in Table 10. We will consider how accurately this symbolism provides insights that aid our understanding of the archaeological evidence estimated to originate during the millennia leading to the glacial maximum of the last Ice Age.

In the previous chapter, we considered the c. 34,600 BC statuette of a lion headed person found in a German cave and the paintings of dangerous animals from the same period on the walls of the Chauvet Cave in the Ardeche region of southern France. The images of revered or dangerous animals suggest that the cultural male focus had shifted from the projection of a hunter's desire for greater hunting skills onto the teeth of powerful carnivores to increasing the belief that they were fine hunters.

This evidence of probable male activities seems at first to be of no direct importance to women. However, the oldest 'Venus' figurines from Europe illustrate how women may have creatively imagined themselves as successful mothers. The 'Venus' figurines found widely distributed from Russia across central and western Europe to the Pyrenees have date estimates of 27,000-22,000 BP, which calibrates to c. 29,000-24,000 BC.[1] The different examples of art from this period illustrating feminine themes suggest that they were part of a belief system that developed into the worship of the Earth Mother goddess.

From Table 16 we see that feminine imagination may have used the keywords from the humanistic Age of Leo. This describes a way in which feminine *self-awareness* could develop whereby women *learned* to use feminine figurines to serve their own goals of motherhood. The use of their *intuition* could have been recognised as the skill that was their greatest *resource*. These figurines may have been the result of the imagination

---

[1] *The Oxford Illustrated History of Prehistoric Europe*, Paul Mellars, p. 72 and *IntCal09 Supplemental Data*, p. 1127-9.

expressing other spiritual keywords of their time, which describe blessings, inspiration and aspirations.

From Table 16 it is estimated that this goddess period spanned some 5,000 years and may have lasted until the last glacial maximum. It is of interest to identify the possible peak in the expansion of feminine symbolism from a fact about the 'Venus' figurines given in the writing of Prof. Paul Mellars. He comments that they appear to be limited to the short period between c. 25,000 to 23,000 BP, which calibrates to c. 28,000 to 26,000 BC.[2] To assess the symbolism aligned with this 2000 year period we refer to Table 16 and see it overlaps with both the humanistic Ages of Cancer and Gemini.

A possible way ideas about the goddess developed during the 2150 year period commencing 29,210 BC is described by the keywords attributed to the humanistic Age of Cancer, (29,210-27,060 BC). It is the quality of the physical *actions* that brought the spiritual *blessings* that motivated this age. Here we have the perfect archetype for the emotional connection between a mother and child. Love in *action* has many expressions including the idea of powerful women coming to the fore that *value* higher standards of mothering within a culture so each new generation of pregnant women is *educated* so they may experience the spiritual *blessings* of motherhood.

When we consider the four phases of Cancer's symbolism spanning 8,600 years we can see these as a continuum of feminine excellence that culminates with blessings. We see from Table 15.1 that the first physical phase of this sidereal sign's symbolism began 35,660 BC with opportunities to develop physical *security* in appropriate ways. This was followed by *nurturing* and *learning* phases before *blessings* were experienced. Cancer's qualities of feeling are symbolised by the Moon and have love as its theme.

The 'Venus' figurines referred to above are a major subject in the title *The Myth of the Goddess,* by the authors Anne Baring

---

[2] Ibid, p. 69 and *IntCal09 Supplemental Data*, p. 1128.

and Jules Cashford. They describe how these images were often made of clay and date from '30,000 to 20,000 BC' and later. These small figures depict various aspects of feminine fertility. Such examples of art could easily be carried for personal use and suggest that a new and widespread cultural priority had arrived; that of women becoming more successful mothers.

Several small statues from Dolni in the Czech Republic are from the period between 25,000 and 20,000 BC. They capture the image of woman as mother and have been interpreted to focus on the birthing process and breast-feeding. A famous Palaeolithic goddess sculpture, the Venus of Brassempouy, was discovered in this place in southwest France. It reveals the fine, delicate features of a female head carefully crafted out of mammoth ivory c. 22,000 BC.

The dates of these artefacts support the idea that many millennia in prehistoric times are associated with the development of a spiritual culture based on feminine values. A prehistoric awareness of the mother's essential purpose for all life on Earth culminated in the belief in an Earth Mother deity. The developments that led to the veneration of the Goddess over nearly nine millennia established feminine values at the heart of various cultures. There is growing academic agreement that the widespread adoption of an Earth Mother culture formed part of a powerful cultural feature of prehistoric life. Her numerous images, created over many thousands of years, are understood to have played some part in the worship of this Earth Mother Goddess. Evidence of her worship is provided by a famous sculpture from the Dordogne area of southwest France that illustrates imagery more sophisticated than that provided by the feminine symbols referred to earlier in this chapter. The complexity of this artwork exhibits dimensions of consciousness beyond the direct concerns of motherhood. Within a Dordogne cave at Laussel a unique bas-relief sculpture dated c. 19,000 BC provides evidence that at that time the Earth

Mother cult had established its Supreme Being, or Godhead. [3] [4]
This image of a pregnant woman, seated and naked is named
the 'Venus of Laussel' and has been interpreted as an
expression of the compassion of the 'Great Cosmic Mother',
holding a bison horn as a symbol of the Moon.   Groups
dedicated to her symbolism may have gathered in the cave to
worship her. A feature of this ochre painted image is that no
facial features survive.   While this could be the result of
vandalism or natural weathering/erosion, it could also have
resulted from a profuse adoration of her, if touched by many
lips and hands.   This image of the Earth Mother godhead
provided a symbol for those in awe of the human dependence
on the female mysteries of fertility, blood and birth magic.
Knowledge of such matters favoured powerful priestesses
leading and teaching other women the wider arts of
motherhood and leadership.

The theme of women making connections to communicate
about shared interests correlates well with symbolism of the
humanistic Age of Gemini, (27,060-24,910 BC). The expression
of this symbolism was extensively considered when we looked
at this same symbolism 25,800 years later when the
precessional cycle's reference point was aligned between 1260
BC and AD 890. In Chapter 9, we saw that this age was
concerned with the growth of monotheism involving Abrahamic
religions of Judaism, Christianity, and Islam. These two
millennia provided many opportunities for influential men to
contribute to these patriarchal religions. From Table 16 the
keywords for this period describe physical *acceptance*
emotional *courage*, mental *answers* and spiritual *inspiration.*
During the previous humanistic Age of Gemini, the evidence of
goddess worship was widely distributed. This feminine
symbolism suggests the possibility of a cultural *acceptance* of

---

[3] *Europe - A History*, by Norman Davis, p. 72.
[4] Rock carvings cannot be directly dated and thus it is possible this image was made c.
27,000 BC and that the dated ochre paint applied much later.

the goddess that gave spiritual *inspiration* together with the *courage* to believe in these *answers* to feminine challenges. In this sentence, we see how related themes suggest the emergence of a matrifocal culture coming into existence.

During the humanistic Age of Taurus, (24,910 BC-22,760 BC) the last glacial maximum (LGM) was present. This time dates when glaciation was at its maximum with sea levels at their lowest. Nearly all ice sheets were at their LGM positions from 26.5 ka (thousands of years) to 19 to 20 ka.[5] However, 'It is known that mountain glaciers and continental ice sheets around the globe reached their respective maximum extent at different times during the last glacial cycle, often well before the global last glacial maximum (LGM; c. 23–19 ka)'.[6] The above data suggests that the LGM was a period during the humanistic Age of Taurus.

The keywords from Table 16 for this humanistic age describe the potential for the *aspirations* of *groups* to take *initiatives* toward those for which they have *compassion*. These keywords suggest different groupings emerged for the successful completion of tasks for which they were suited. With communities having to survive during this time of maximum glaciation, life's two fundamental criteria became most testing, the availability of food, and people's safety. It is possible the alignment of this sidereal sign with the LGM is most fortuitous since its symbolism correlates well with the function of the Earth Mother. The qualities that were introduced into consciousness during the continuum of Taurus are included in Table 16. These begin with an appreciation that during the humanistic Age of Leo all physical *resources* are synonymous with the power of the Earth Mother. During the humanistic Age of Cancer, the *value* of these resources to support life is recognised. During the humanistic Age of Gemini, these

[5] *Science 7,* August 2009: Vol. 325 no. 5941 pp. 710-714
[6] *Earth-Science Reviews Volume 125*, October 2013, Pages 171–198

qualities were providing the *answers* needed. The last keyword from this continuum recognises how the Earth Mother's *aspiration* is to nurture all expressions of life

For the humanistic Age of Aries, (22,760 BC-20,610 BC) the *groups* that developed during the previous age have the potential to become more proficient and increase their success. Hunters may have attracted *friends* with the *will* to *contemplate* which of their hunting techniques *established* the greatest success. The interests of women are also likely to have been more successful when they formed groups to help new mothers and their children. Once the advantages of people contributing as a group were recognised and this became part of the culture, it is probable that groups formed in different locations with specialised skills that used locally available resources. For example, a locality rich in high quality flint would favour local quarrying and tool making skills for their region's needs.

As the cold from the last Ice Age approached its maximum and became more penetrating, hunters had to develop their skills. One idea for this that has fascinated palaeontologists since their discovery was that the cave art was used with hunting in mind. The archaeological evidence from many different Upper Palaeolithic art sites contributes clues about how the consciousness of our ancestors developed during this period.

The cave art produced during the coldest times of the last Ice Age was not a new phenomenon. The drawings in the Chauvet caves predate the Spanish and French cave art works by some 10,000 years. However, the spiritual processes the c. 21,000 BC hunters experienced and the results they achieved may represent an advance of earlier techniques. It is possible the earlier focus was on the ability of the individual hunter, whilst the hunting being considered here concerns the ability of the group to hunt. In such cases, the hunter's initiative naturally engages the human *will* to focus or *establish* some form of connection with the group and the species of game being hunted. The group concentration on a painting of the animals hunted would be one method to achieve this.

Around the LGM, the French Dordogne as a centre of the cave art makes it a likely candidate for being an important place for pioneering human progress at this time. A feature of some caves in which Palaeolithic art has been discovered is that their locations are deep underground. Artists had to crawl hundreds of metres through small tunnels to reach many of these galleries. It is striking that the extensive passages that had to be negotiated to reach the paintings resemble the fallopian tubes that carry the fertilised egg into a woman's womb. With a burning torch to provide some light painters expressed their creative genius within these dark and silent wombs. Groups may have used these paintings as a sacred destination to provide a mental focus to purify their collective will.

The keywords for this humanistic Age of Aries, to physically *establish*, emotional *friendship*, mental *contemplation* and the spiritual *will,* may be considered as qualities that provide opportunities for the skills of hunters to reach a peak of excellence. This possibility complements a theme introduced earlier in this chapter where we saw how the four phases of the continuum of Cancer's symbolism created the potential of a continuum of excellence for motherhood. When we look at the continuum of physical, emotional, mental and spiritual keywords for the symbolism of Aries, and consider the qualities most suited to the successful Stone Age man – *action, courage, initiative,* and *will* all contribute qualities essential for success.

The cave art created around this time includes the world's richest examples from southwest France and northern Spain. Reasons for its creation are considered in the book *Cave Art*, where Paul G. Bahn describes many of the theories suggested by specialists in this field. The one that best fits into the scenario being developed here draws on research obtained from the study of Australian Aborigines. Part of the Aboriginal tradition was to 'perform ceremonies in order to multiply the numbers of animals, and for this purpose they painted likenesses of these species on rocks.' This technique describes the creative visualisation of ample game to hunt.

To see if these ideas about cave art during the humanistic Age of Pisces, (20,610 BC-18,460 BC) correlate with the aligned symbolism we will look at the physical, emotional, mental and spiritual keywords from Table 16 to see if they describe a process visualisation. The LGM challenge to find food demanded the discovery of new *opportunities.* It is possible that shaman with the spiritual capacity to *transcend* normal ego limitations, communicated with the intended prey animal. It is possible such a figure would have the mental *intelligence* and *authority* to guide the hunters. This possibility fits with a theory concerning the use of cave art for magical hunting practices.

17

The Ending of the last Ice Age

During the last eight millennia of the Upper Palaeolithic and the start of the Mesolithic, the warming environment imposed a continual demand for cultural readjustments. Insights about these changes have been gleaned from a consideration of the archaeological evidence from different periods and regions. Much of the evidence selected for this study comes from the writings of Prof. Steven Mithen from Reading University in England. In particular, his chapter on the Mesolithic Age in *The Oxford Illustrated History of Prehistoric Europe* and his 2003 book *After the Ice – A Global History*.

A change that took place over these millennia was the melting of ice sheets up to three kilometres thick and the rise in sea level. Most of the world's former coastlines are 120 metres below the present sea level. Exceptions to this are regions where the ice was very thick. This weight of ice on the northern lands caused them to float lower on the Earth's molten core than the unburdened landmasses. In Scandinavia for example, where the ice was thickest, the land when unburdened floated higher on the molten core. The shoreline at the time of the last glacial maximum is now over 250 metres above today's sea level.[1] In Norway and Sweden, it took many thousands of years for the Earth's crust to rebound causing the valleys carved out by the earlier glaciers to be reborn as its famous fjords.

In coastal regions, a critical theme of the warming climate was the varying speed at which the sea level rose. One reason for this was that the rate at which the environment warmed was not constant. The second reason concerns how huge volumes of ice melt were often held in vast inland lakes by ice

---

[1] Steven J. Mithen, *The Mesolithic, Age Prehistoric Europe,* p.81.

dams. When these dams broke vast quantities of water were released. The dramatic erosion of land in several western American states is one illustration of the power of meltwater to carve channels to the sea. Particular periods have been identified when meltwater reached the oceans in such surges. The size of these meltwater pulses (MWP) varied greatly. Details of some pulses included below are taken from NASA, Goddard Institute of Space Studies.[2] From the G.I.S.S. Generalized Curve of the rate of Sea Level Rise since the last Ice Age, the following meltwater pulses have been included in Table 17 below to suit the BC date scale being used.

Table 17

New qualities of consciousness introduced in Western Europe during the ending of the Upper Palaeolithic and the start of the Mesolithic

Technology	Solutrain	Magdelanian		Azilian	
meltwater pulses	M-W-P 1AO	M--W--P 1A		M-----W-----P 1B	
climate themes	←LGM	I Bølling-Allerød I	Younger Dryas I	Holocene climate	
Sidereal signs **of the** **Zodiac**	Humanistic Ages of the late Upper Palaeolithic and Mesolithic				
	Aquarius from 18,460 BC	Capricorn from 16,310 BC	Sagittarius from 14,160 BC	Scorpio from 12,010 BC	Libra From 9,860 BC
Aquarius	*ideal*				
Capricorn	*knowledge*	*sacrosanct*			
Sagittarius	*honesty*	*truth*	*belief*		
Scorpio	*release*	*share*	*transform*	*rebirth*	
Libra		*evaluate*	*inter-depend*	*agreement*	*goodwill*
Virgo			*serve*	*care*	*discriminate*
Leo				*identity*	*confidence*
Cancer					*security*

[2] Vivian Gornitz, *The Great Ice Meltdown and Rising Seas: Lessons for Tomorrow* from NASA Goddard ISS.

The 2nd row of Table 17 shows some major environmental milestones within the context of their respective humanistic ages.

MWP-1A0, c. 19,600 to 18,800 years ago, [17,600-16,800 BC]
MWP-1A, 14,600 to 13,800 years ago, [12,600-11,800 BC]
MWP-1B, 11,000-9,000 years ago, [9,500-7,000 BC]

When we look at the 2150 year period of the humanistic Age of Aquarius, (18,460 BC-16,310 BC) we recognise that the first pulse of meltwater (pulse-1A0) after the last glacial maximum meltwater pulse raised the sea level by some 10 metres.[3] This caused the flooding of many coastal regions. Around this time, the migration routes of many animals changed. The hunters and their families, which depended on them for food, had to change their lives as necessary.

The animal bones found in ancient settlements reveal that many groups from Europe were dependent on reindeer. South-western France and the Cambrian mountains of northern Spain were probably the most highly populated region of Europe and has proved to be a rich source of information about communities living through the challenges presented as the Ice Age melted away. We can expect that these changes created new conditions where hunting traditions were forsaken, as sacred seasonal hunting sites on coastal plains were flooded.

It is likely that the cultural changes imposed did not detract from an emphasis on successful hunting. The necessary visualisation skills possibly exercised by shaman may have become a cultural necessity for all hunters. During a time when the hunting environment was changing, the keywords from Table 17 may capture descriptions of the challenges addressed. At the spiritual level, the theme of the hunt may have been to visualise an *ideal* result together with an abundance of prey to hunt. Mental qualities use the *knowledge* gained as the skills of hunters were honed. Emotional *honesty* was necessary when

---

[3] Vivian Gornitz, *The Great Ice Meltdown and Rising Seas: Lessons for Tomorrow* from NASA Goddard ISS.

recognising that traditional hunting practices were obsolete. These had to be *released* as new lessons and methods for success were adopted.

Many archaeologists support the theory that the cave art of this period was created to serve the success of the hunter. Paintings from the famous Lascaux cave in the Dordogne's Vezere valley is one of the world's richest sites for examples of cave art. The walls and ceilings of this site's caves were covered with nearly 2,000 images; most of them illustrate the prey animals people depended on. The famous Great Hall of Bulls depicts bulls, equines and stags. Nearby caves, have paintings of ibex, bison and cattle. Few lions, bears and wolves appear. Lascaux cave art dates to c. 18,000 BC. [4] This site together with the northern Spanish cave of Altamira may have been cultural centres for survival rituals. These could have served communities so that hunters provided a reliable supply of food from healthy stocks of game animals.

When the humanistic age of Capricorn (16,310 BC-14,160 BC) began the gradual warming of the climate was slowing down and the use of cave art possibly reached its most prolific expression. The keywords for this humanistic age suggest that the cultural significance of this cave art was *sacrosanct.* If the art served a practical purpose for successful hunting practices, it is possible the artists were revered figures, possibly spiritual leaders. If survival during much of the Ice Age depended on this culture, the need to *evaluate* its practicality, as experienced by the community that *shared* the results, needed to be *truthful.*

Communities that followed a culture that focused on this cave art had to learn to live in the advancing forests encouraged by the new climate. It is possible a cultural alternative arose when the availability of game was deemed more sacred than the cave art. This priority became a new challenge since the amount of game available from forested environments is substantially less than that available from the open tundra. This reduced the

---

[4] This BC date is calibrated from the 17,000 BP date given in *Prehistoric Europe*, p. 69.

size and number of local populations supported in this way.[5] These forest dwelling groups left very modest remains compared to those of earlier times.

When hunting migrating animals was *sacrosanct* to people, they more likely had access to adequate quantities of food. The other keywords for this period suggest that a vital *truth* would be to continually *evaluate* and *share* with other groups the knowledge of the new, more northern routes taken by migrating animals. This movement of the reindeer took place as they followed the ice frontier northwards. This second scenario describes how keeping with the reindeer could be a cultural theme as *sacrosanct* as the cave art they left behind. The discovery of mobiliary art statuettes of hunted animals suggests a spiritual interest in game animals was expressed c.14,000 BP (15,000 BC [6]) in regions where cave art did not exist.[7] Prof. Steven Mithen's book *After the Ice* describes how around 15,000 BC settlers moved north of the Loire's catchment area. By 14,500 BC, they reached the Rhine, which led their descendants and other groups downstream towards Belgium.[8]

By the start of the humanistic age of Sagittarius, (14,160 BC-12,010 BC) the ice had retreated from southern England and northern Germany. The warmth from the sudden Bølling oscillation and Allerød period accelerated the environmental dynamics of this age. See Fig 17 below.

Some of opportunities these provided for the pioneers are described in Mithen's book.[9] Close to the frozen landscape, the barren tundra began to be transformed as moss and lichens established new sources of food that attracted the herds of migrating reindeer. Complex processes followed each incremental step as the ice frontier slowly retreated north. From the cores of lake, sediments laid down over many millennia the successive layers show the pollen of the

[5] Steven Mithen, *The Mesolithic, Prehistoric Europe*, p. 75.
[6] IntCal09, P.J. Reimer, p. 1122
[7] Paul Mellars, *The Upper Palaeolithic Revolution, Prehistoric Europe*, p. 68.
[8] Steven Mithen, *After the Ice*, p.120.
[9] Ibid, p.113.

increasing variety of grasses, shrubs and trees. These records provide evidence of the vegetation that developed in different regions of Europe. The colonisation of landscapes by a widening variety of plants, birds and animals provided a springboard for the more rapid changes when the temperature rose very rapidly with the sudden arrival of the Bølling oscillation.

Estimate of Climate Influences based on GRIP Ice Core data

Fig 17

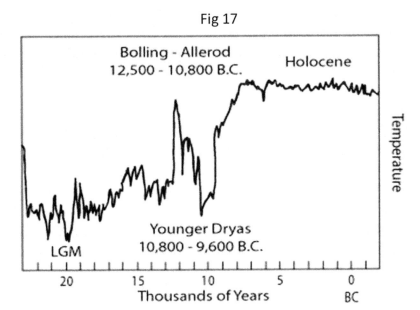

This warming greatly accelerated the rate at which many environments were transformed. The warmer and wetter climate released more tracts of land to become virgin forests rich in flora and fauna.

These resources attracted hunter/gatherers to settle in these regions. The survival of those people that migrated northwards with the reindeer likely demanded acute attention to the changing dangers and opportunities that came with the melting ice. The keywords from Table 17 suggest that the warmer climate encouraged new *belief* about what was now possible in their new environment. Continuing successes could have aided an emotional *inter-dependence* between groups *serving* each other by sharing information about local opportunities and dangers. Such experiences would likely promote this *interdependence,* and *transform* the culture and shape the early Mesolithic.

With plenty of resources for groups to share, their cultures had the potential to express *care* and consideration for each other as well as their local resources. The idea of reaching mutually favourable *agreements* is also suggested by the emotional and mental keywords of the humanistic age of Scorpio (12,010 BC to 9,860 BC). This period was the last of the continuum of physical, emotional, mental and spiritual phases spanning 8,600 years that culminated with the rebirth of human life living in harmony with its environment. The *identity* of the Ice Age hunter/gatherer was *reborn* as they became the masters of the more benign environment in which they lived.

Such behavioural characteristics would have doubtless served communities as the coldest conditions of the last Ice Age returned c. 10,800 BC, suddenly cooling the warm climate after the Allerød period. This 1,300 year mini Ice Age called the Younger Dryas ended c. 9,600 BC, when the rising temperature introduced the current Holocene Age. This change of climate determined the limit of the northward retreat of the ice tundra.

Flooding was a continual threat during the humanistic age of Libra, (9,860 BC-7,710 BC). Table 17 shows a meltwater pulse lasting from c. 9,500 BC to c.7000 BC. These dangerous times resonate with the physical keyword for this age, *security*. The efforts people had to make to find somewhere safe to live can only be imagined together with the *goodwill* that individuals and groups extended to those in danger of drowning. With these challenging conditions, primitive boats could have been vital to save lives. People able to *serve* in this way had to be justly *confident* in their ability to *discriminate* between safe and dangerous conditions when venturing onto the water.

In parallel with the European Mesolithic progress described above, the early advances in farming from the Near East were also made during the period just before the Younger Dryas. The above use of keywords suggesting cultural descriptions during the Humanistic Age of Scorpio are also suitable for suggesting the early advances in farming in the Near East. The *identity* of suitable food seeds depended on their *careful* selection. Non-shattering grain species were *agreeable* for the practical harvesting necessary for the *rebirth* of viable local cultures.

The Neolithic Revolution in the Near East introduced farming to support growing populations. In Europe communities continued to flourish by hunting and gathering. Many millennia were to pass before farming was needed across this region.

18

Modern Correlations

Cyclical Themes.

We have explored how the continuum of our emerging consciousness has been motivated by physical, emotional, mental and spiritual initiatives over many millennia. The cyclical nature of this process is illustrated by the precessional cycles considered. Evidence of this cycle starting twice comes from the cultural rebirths that began during the humanistic ages of Scorpio. Firstly, the c. 38,000 BC starts of the Upper Palaeolithic Revolution and secondly the c. 12,000 BC beginning of the Neolithic Revolution: span the Great Year's precessional cycle's duration of 25,800 years.

The first of these rebirths came with the imagination being used to design the stone tools then invented. With these new tools, hunters had better weapons to kill and butcher the game they depended on. The second rebirth came with the start of the Neolithic, as hunters became farmers dependent on the seeds of plants suitable for providing new staple foods.

The next cyclical theme is the correlation between the culture that introduced the worship of the Earth Mother goddess and that, which fostered the beginning of monotheism. In chapter 16, we considered how a matrifocal religion stemmed from the development of a feminine culture. This is identified by the introduction of numerous Venus figurines, c. 27,500 BC. [1] These artefacts are evidence of the growth of a new culture, initially focused on successful motherhood. It developed to include many other aspects of the feminine. It culminated in the worship of a supreme deity, the Earth Mother. The period during which the worship of this goddess developed

[1] See p. 169 Table 15.3.

suggests that several millennia passed before this deity became an integral part of European culture. Some 25,800 years later, an Egyptian culture was the first to adopt monotheistic spiritual beliefs for a short time. This cultural initiative was later followed c. 1260 BC by beliefs and the worship of the Hebrew God represented by their leader Moses.

Current Precessional Alignments.

We will now look for other themes emerging during the precessional cycle. To do this we will precession in another way. Earlier it was useful when we found that the precessional 'summer solstice' aligned with the date of 1260 BC, early in the reign of Rameses II. This followed the heretic Pharaoh Akhenaten's unsuccessful attempt to introduce monotheism throughout Egypt.

We can use the date of 1260 BC for another purpose. From this 'summer solstice' point in the precessional cycle, we can find how far the precessional position of the spring equinox has moved since this date.

To do this we will first determine how many years have passed between the precessional 'summer solstice' and 2015. From this, we will be able to calculate how many degrees of precession the Earth has moved through since this solstice. From the date of 1260 BC to the date of 2015, we find that 3,275 years have passed.

With our next calculation, we want to find out how many degrees our reference point has moved through during these 3,275 years.

This number is found with two calculations. Initially we need to calculate what fraction these 3,275 years are of the 25,800 year precessional cycle's 360 degrees.

Dividing the number of years 3,275 by 25,800 gives the figure of 0.1269379.

This is a little more than an eighth (0.125) of the cycle. 360 of degrees divided by 8 = 45 degrees.

The figure of 0.1269379 is then multiplied by 360 degrees to calculate how many degrees the reference point moved during these 3,275 years.

The answer of 45.697674 degrees, or put another way, 45 degrees 41 minutes and 51 seconds.

This tells us that the spring equinox has moved from its 1260 BC precessional 'summer solstice' place, to its 2015 position, a little more than 45 degrees.

This movement through the thirty degrees of Gemini continued until it passed the middle of the sidereal sign of Taurus. From the above arc, we calculate that the 2015 spring equinox position is 15 degrees 41 minutes and 51 seconds from the end of Taurus, or 14 degrees 18 minutes and 09 seconds from its seasonal start.

We now need to know if this position in Taurus is in an alignment with any significant features of the zodiac's symbolism. Interestingly, this position of the spring equinox is just past an exact conjunction with the alpha star of this sidereal sign, Aldebaran. This star's position defines the centre of the sidereal sign of Taurus as 15 degrees 00 minutes 00 seconds in the Babylonian sidereal zodiac for the epoch 1950.[2] This star is one of the most useful stars of the zodiac since it is the brightest point on the annual path of the Sun around the heavens. It was originally chosen to denote this cycle's start.

This usefulness is supported by the presence of the star Antares positioned almost in the centre of its sidereal sign of Scorpio at 14 degrees 58 minutes 25 seconds. This position is nearly exactly opposite the star Aldebaran in Taurus. This creates a natural reference axis dividing the zodiac into two practically identical halves. A feature of this relationship is that as one of the stars sets the other is rising over the opposite horizon. Astronomical observatories on islands with suitable sea level horizons were practical early sites for astronomers.

---

[2] Every fifty years a precise time is used to define the celestial coordinates or orbital elements of astronomical phenomenon. This practice has been adopted because many influences change the positions and speeds of celestial objects. As Aldebaran's motion is 3.62 seconds per century the 1950 position is sufficiently accurate for our purposes. Powell and Treadgold, *The Sidereal Zodiac,* p. 45.

These two stars are included in the ancient Royal Stars of Persia since they were recognised by this culture for their powerful symbolism for success. Aldebaran was called the Watcher of the East and considered the guardian of the spring equinox.[3] Antares, the Watcher of the West was associated with the setting Sun. In Egypt, it represented a form of death.[4] This applies to the autumn equinox and the coming winter as well as the successful completion of demanding tasks. Their different symbolism developed from their association with these two seasons of the year. These positions in the annual cycle provide some ideas about the Sun's creativity to consider. The benefits symbolised by Aldebaran concern the successful expression of life's creativity. The welcomed warmth of springtime heralds the awakening of the environment's capacities to burst into life. The new leaves of plants, the beauty of blossoms and flowers carry the promise of the coming season's bounty of new supplies of food for all. Such success is not automatic. Many conditions are necessary if this promise is to be kept. Gardeners are well aware of the threats easily able to destroy or limit their wishes. Late frosts, strong winds and little rain are just a few of the dangers to guard against if the full potential of invested effort is to be realised. Extra work is usually vital to minimise the unwanted influences that threaten failure.

When we consider the symbolism of Antares and its association with death and decay, we recognise that the fertility of the soil is gradually exhausted as plants grow. This degradation of the soil, if repeated year after year without its fertility being replenished, would eventually leave the land barren and lifeless. The annual loss of the land's fertility illustrates environmental changes that correspond with the expression of the symbolism of Antares. The action of the frosts and rains of autumn are vital for aiding organisms to reduce rotting waste vegetation down to its constituent elements. The

---

[3] Bernadette Brady, *Brady's Book of Fixed Stars*, p. 233.

[4] Ibid p. 288.

success symbolised by Aldebaran is that the appropriate replenishment does take place.

Now that we have some basic ideas about the inter-dependence symbolized by these two stars, we will see if these ideas can be related to historical events aligned with the time when the precessional 'spring equinox' was conjunct Aldebaran. We want to estimate a time when this symbolism may have been displayed on a global scale. To help us calculate this we need to do some more calculations to determine the date when the spring equinox was aligned with Aldebaran's position at 15 degree 00 minutes and 00 seconds of Taurus.

We start by calculating the annual arc of precessional movement. To do this we divide 360 degrees by 25,800 years. This tells us that in one year precession moves the position of the spring equinox along the path of the Sun through an angle of 50.23 seconds of arc.

We know from earlier calculations that the 2015 spring equinox reference point had moved 41 minute and 51 seconds beyond an exact alignment with Aldebaran. For the purpose of the next calculation, we will convert this arc into geometric seconds; 41 minutes multiplied by 60 plus 51 gives us the answer of 2511 seconds.

When we divide this angle by 50.23 seconds, (the amount precession moves in one year) we get the answer of 49.9900 years, virtually 50 years. This period when subtracted from 2015 gives us an estimated alignment date of spring 1965. See Fig 18 for the positions of the spring equinox in the sidereal zodiac for 1962, 1260 BC and10,900 BC.

Fig 18 below illustrates our alignment with the stellar opposition between Antares and Aldebaran within the cycle of precession. It took 12,900 years, half a cycle for the spring equinox to move around from Antares to Aldebaran.

Fig 18.

The Great Year – The Precession of the Spring Equinox
Clockwise motion

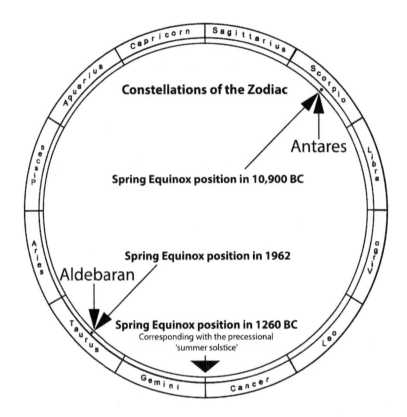

Does the estimated date of the alignment between Aldebaran and Antares identify a time when many major agendas, which attracted global attention during the 1960's revolutions, became clearly identified?

Many events of this decade display the challenges still present in modern cultures. Central to these is still the emergence of repressed cultural energies demanding greater *compassion*. We will look at other qualities and dimensions of experience to see how they continue to describe vital themes for our age.

We will first consider the early *aspirations* of Chinese communism to transform an agrarian culture into a country with modern national industries. The mandatory collectivization of agriculture spanning the period from 1958 to 1961 caused much suffering for large *groups* of people. These huge failures illustrate a lack of *compassion* inherent in the national plans to *initiate* their country's Great Leap Forward.

In subsequent conferences in 1960 and May 1962, the Chinese government studied the negative effects of this Great Leap. The work of Mao Tse-tung successfully contributed to expose corruption in his country's Communist Party. Within a few years, his Little Red Book became a symbol for the Cultural Revolution that swept China towards greater achievements.

The Chinese agricultural failures were not the only man-made disasters with dramatic environmental consequences. The role of the goddess to succour life was betrayed in other ways. The Vietnam War included the infamous use of Agent Orange to defoliate Vietnamese jungles. This environmental issue alongside the testing of nuclear weapons devalued ecological concerns. In 1962, Carson's book *Silent Spring* led to the start of the USA's Environmental Protection Agency.

The theme of the aspiring feminine is symbolized by some more famous examples during 1962. These suggest the earlier estimate of 1965 be reviewed. The birth control pill was available on the National Health Service in Britain in 1962. In this year, Betty Frieden was finishing her book *The Feminine Mystique*. It initiated a new wave of feminism with its descriptions of the American housewife's dissatisfaction with the narrow social roles enforced upon them by their countries culture.

The international issue of this year concerned the pinnacle of a major ideological post World War II struggle. The map of strategic global power had polarised into two camps: the *aspirations* of Russian communism challenged the spread of American capitalism. The tensions between these two powers reached their most dangerous in 1962 when US defense forces went to the highest state of alert. A global nuclear war was narrowly avoided during the Cuban missile crisis when America and Russia came close to the brink of mutual annihilation. The size of our planet had imposed its limits on national ambitions. Nuclear disarmament entered into the political debate. The reality of us all having to live together on Earth continues to be expressed in increasingly complex protracted political negotiations.

More themes erupted in 1962 when many young people displayed their *aspirations* to be free from limiting traditions. In America, the civil rights campaign and anti-Vietnam War protesters were joined by the hippy counter-culture of flower power and sexual liberation. Protestors placing roses down the barrels of soldiers' rifles expressed the abnormal friendliness of the period. The songwriter and singer Bob Dylan was the reluctant leader of a multi-faceted revolution sweeping the West. One of its anthems, the song *'Blowing in the Wind'*, was written by Dylan in 1962.

Another musical milestone was created in the same year when Brian Epstein negotiated with EMI a recording contract for The Beatles, the famous Liverpool pop group. The global sales of their recordings were over half a billion units. This is a measure of how well their musical and song-writing skills expressed new energies able to unite teenagers across the world. A landmark within the entertainment industry during 1962 was the launch of the BBC television satire programme *'That Was the Week That Was'*. It ridiculed any dimension of the UK establishment that failed to meet the standards of excellence that they purported to embody.

The civil rights worker James H. Meredith, an African American formerly in the Army Air Force applied for a place at the all-white University of Mississippi in 1962. Racial intolerance was expressed when the registrar discovered his colour and withdrew his admission. Meredith filed a suit alleging discrimination and despite receiving a favourable decision from the U.S. Supreme Court, Meredith found his entrance to classes blocked; rioting erupted. The action of the Attorney General to send U.S. Marshals to the scene was followed by President Kennedy ordering a military escort from the National Guard to accompany Meredith for his enrolment during October 1962.

When we consider the mental realm, 1962 saw the *initiative* concerning the expansion of global communications. Telstar, the world's first commercial communications satellite, was launched successfully into orbital operation.

During 1962, Pope John XXIII convened the Vatican 2 Council. These ecumenical *aspirations* were thwarted by the strength of the conservative grip on Catholicism. When asked why the Council was needed he reportedly opened a window and said that he wanted to throw open the windows of the Church so that 'we can see out and the people can see in'.

The above events of 1962 established the agendas for several of the cultural revolutions associated with this decade. They suggest that this year be selected as the time when cultures began to accept the changes needed to establish greater cultural honesty from both the establishment and individuals alike. Using 1962 for the alignment between the spring equinox and Aldebaran replaces the 1965 estimate.

We now need to see if the cultural dynamics that appeared the 1960s are accurately described as expressions of the symbolism of Aldebaran and Antares. Earlier we considered their symbolism in the context of the annual cycle of the seasons where they represent the masters of spring and autumn respectively. To look for their expression in human cultures we recognise the different skills a healthy society needs to nurture its continuing success. The historical importance of these in ancient Egypt, Persia and Greece is included in Bernadette Brady title, *Brady's Book of Fixed Stars*. It describes how Aldebaran symbolised greatness with 'success linked to integrity of morals and objectives' that require 'purity of thoughts and feelings'.[5] Here we see the potential for the expression of an unimpeachable sense of self-worth that flowers as the portrayal of personal greatness. Such stature displays the individual's values; how they are immune from compromise by the mediocre standards of socially acceptable behaviour.

For ancient Egyptians the success symbolised by Antares was recognised in a particularly vital dynamic. It depended on energies represented by the heart of the scorpion.[6] Its symbolism describes the extremes that had to be reached or demanded, if success was to be achieved. If a cultural ideal was to become reality, some sort of death, transformation or rebirth of old ways had to take place. Real fulfilment depended on the cultivation of skills that bring the death of personal qualities and spheres of incompetence that risked failure. This result may take many forms. Each autumn waste vegetation rots away to purify the resources needed for future plants.

Turning now to the modern cultural expression of the imagery of Aldebaran, we know how success in modern cultures is often created by *aspirations* running through the lives of people, groups and governments alike. This feature correlates well with the symbolism of Aldebaran since this actual star is the stellar light that defines the centre of this sidereal sign, the bull's eye. The bull's eye is the term used to describe every objective, goal or target that focuses our attention. The success to achieve each personal objective is the symbolic bull's eye towards which we strive. The changes always necessary for unprecedented success require the qualities symbolised by Antares as well. These describe the abilities needed to purify disruptive energies; they have to die, transform or be reborn as the strengths needed for success.

During the 1960s, the success symbolised by Aldebaran was seen by many young people to be reserved for the self-serving

---

[5] Bernadette Brady, *Brady's Book of Fixed Stars*, p. 233.

[6] Bernadette Brady, *Brady's Book of Fixed Stars*, p. 287.

standards of behaviour the status quo had institutionalised. A generation of young adults appreciated that the widespread corruption that the world had long accepted would not be tolerated.

A major agenda of transformation was demanded to ensure that ethical behaviour eventually becomes the rule rather than the exception. Vocal minorities were outspoken in their criticism. This blunt candour shocked the countless factions of the establishment formerly insulated from such criticism by misplaced deference.

Cultural themes that propagate cultural separation, which are ripe for reform now, are so regularly headlined that few need listing here. During the last fifty years, many expressions of extremism have erupted and continue to demand increasing global cooperation. If this progress exposes underlying issues of intolerance in need of attention, it is likely unresolved issues of the 1960's are still being activated.

While much cultural healing in still necessary, we can recognise the many fields of endeavour that do embody the highest ethical standards. The international aspirations to protect our natural environments and their species are mirrored by the care and compassion many citizens give to their personal spheres of concern. This experience of greater connectedness and the experience of non-separation are growing despite the greater separation exclusivity breeds by design.

Clearly, the aspirations of various groups competing for power continue to be a real threat to the individual and our planet. The various priorities given by different groups to the degree of *compassion* they express continue to delay our civilized advance. Has the precessional cycle any way of confirming this observation?

Cyclical Alignments

The revolutions of the 1960s are recognised to have been of

global significance. In view of this, it is of interest to assess if failures to address its continuing agendas have implications for the integrity of the complete precessional cycle. For clues about this vital matter, our attention will focus on our current place in the cycle. Its context is provided by several pivotal themes of consciousness earlier emerging in the cycle.

Firstly, the c 38,000 BC date of the Upper Palaeolithic Revolution is linked by 25,800 years with that of the Neolithic's 10,900 BC date. Secondly, c. 24,000 BC dates of the worship of the Earth Mother goddess agenda and is linked by 25,800 years with our current difficulties with monotheism and ecological aspirations. These two themes are cyclically related by the second theme beginning exactly halfway through the first.

Fig 17 on page 190 estimates the pattern of temperature changes during the last Great Year. The sudden rise in temperature c. 12,500 BC and the following warm period, the Bølling-Allerød, suggests a fascinating alignment. The intrigue arises because the date of the end of this warm period, c. 10,900 BC is 12,900 years ago and directly opposite the 1960's precessional position of the spring equinox: see Fig 18. The 12,900 years between these dates fit nicely with the c. 12,900 year period between the Upper Palaeolithic Revolution and the peak of the worship of the Earth Mother goddess. These half cycles link up two precessional cycles.

How can these global changes of the last precessional cycle possibly help us understand our modern challenges? This pattern suggests opportunities to assess if our current global challenges relate to activities that are not in harmony with the process described by these precessional cycles.

When we look on the precessional symbolism aligned with 10,900 BC, we focus on that associated with the star Antares. Its symbolism describes how success arises from agendas concerning death, transformation and rebirth. The successful renewal of needed resources is the primary objective. In the environment, we can associate this need with the resources vital for the potential encoded in seeds to be physically expressed as plants grow. In human affairs, this principal corresponds with the potential to cooperate with the opportunities our planet provides to increase our skills in physically expressing our spiritual creativity.

A Common Structural feature of each Humanistic Age.

In an attempt to assess how well we are doing, it will be useful to look at some patterns of change that took place during past humanistic ages. After the start of each age, the early success of each spiritual initiative was followed by difficulties that limited or delayed their widespread adoption.

Where we have historical evidence, we will trace these initiatives and recognise that the challenges were overcome many centuries later. We will begin by looking at the famous pinnacles of achievement reached in the Egyptian Old Kingdom, Israel's court of King Solomon and the spiritual possibilities the Renaissance introduced for the independent mind.

The works of the first Egyptian kings and pharaohs provide the first recorded example of this pattern. Their spiritual initiative concerned the divine right of kings to rule over the Egyptian nation. Their early successes were not sufficient to ensure their Old Kingdom was eternal. It faltered before collapsing after the Nile failed to flood for several years.

Many centuries past before this theme of divine rule was developed further at the start of the next age. The idea of One supreme god responsible for all Creation became the Hebrew religion when Moses introduced the first of the Abrahamic faiths. The successes of the nation of Israel united and ruled by King David, the wise reign of Solomon and the building of his Temple did not last. The Assyrians exiled the ten lost tribes of northern Israel to Persia c. 720 BC. The two remaining tribes remained in their southern lands.

At the start of the next age, after c. 800 Charlemagne created a Christian Empire between the Atlantic and the Danube, he initiated the Carolingian Renaissance. The resulting education of the clergy enhanced the growth of monasteries in Europe during the early Middle Ages. The keywords for this age *aspiration, initiative, compassion* and *groups* were being expressed in monastic activities by its AD 880 start. This example illustrates the challenges still evident if a new type of consciousness is to become a global culture.

The benefits monasteries introduced into their surroundings included the monks being charged with doing God's Will by caring for the poor. In monasteries the Almoner was responsible for fulfilling this biblically prescribed duty; distributing alms. This at least was one sphere of Church doctrine, in which a monk's *aspiration* was very likely to be in alignment with his higher will, conscience and *compassion*. The wish to serve in this way echoes priorities evident when the Earth Mother was the cultural focus of our emerging consciousness. The divinity of this feminine polarity displays priorities long eclipsed by patriarchal *aspirations*.

The education of monks became a vital part of monastic life. Their curriculums expanded after the scholastic fruits of Islam's golden period became available to them and included many works from ancient Greece, which Islamic scholars had earlier saved and studied. These ancient scriptural and philosophical works nurtured a reasoning of their faith in Life's unity.

Monasteries became regional centres of *initiative*. Through their welfare work with hospitals, education, books, art and architecture, these communities thrived. Their extensive lands expanded agricultural activities; their architecture still dominates many landscapes, their *compassion* for the poor established the first hospitals. The range of skills they introduced to their localities provided models that preceded the eventual widespread secular provision of these vital services.[7]

The benefits the monasteries gained from education

---

[7] Dr Janina Ramaries, *Saints and Sinners*, BBC TV Series by Oxford Film and Television Production, 2015.

attracted secular participation. As this increased, secular *aspiration* and *initiatives* expanded and gradually gave birth to the Renaissance. For many centuries, the traditions of the Church had dominated European culture. After Roman laws were no longer enforced, Church doctrines became the tools for cultural power. The authority behind the spiritual traditions of the Church was debased by corruption and the abuse of power. The knowledge and wisdom from classical civilisations provided attractive alternatives. The demand for orthodox worship was undermined by the personal experience of following one's conscience. This humanistic movement was led by German demands for spiritual plurality. Their mystical quest included the individual's potential to experience their own self-awareness of the Divine. This freedom was aided after the Bible was translated into the aspirant's native language.

The revolution printed books introduced extended the study and discoveries made by science. The orthodox teachings of the Church about Creation were contradicted by a succession of famous astronomical discoveries made by Copernicus, Galileo and Kepler. The greater emphasis placed on reason undermined the need for faith. Education expanded with its secular success in achieving personal, national and international ambitions. The English Crown rejected the worldly power held by the Church.

After the dissolution of the monasteries, centuries passed before any secular cultural responsibilities emerged to embrace this need for welfare and civil works. It is likely the repercussions of this ethical absence are still with us.

The greater understanding of the laws of nature fostered the growth of knowledge that introduced the Scientific Revolution. The power of this a new cultural dynamic led to the debates of the Enlightenment. Philosophers believed that the scientific advances would introduce changes to help the poor.

The later Industrial Revolution began when water wheels focused on the weaving of cloth in factories. Cottage industries dependent on rural landowners declined. The workers needed for the new factories were attracted away from their former serfdom dependence. The first Quaker factories provided good living and working conditions for their employees.

The use of the first steam engines to pump water out of mines replaced the horses that pulled the wagons of mined coal. Railways quickly expanded and transported other raw materials to factories and delivered manufactured goods. The potential for greater wealth after the start of the Industrial Revolution created an unstoppable momentum as more factories were built. The sweet success supplied from the Caribbean was not enjoyed at its cane plantation workers.

The eloquent warnings from the Romantics about this rush towards our continuing consumer delusion were generally ignored. Science introduced the technology to build the ships and weapons that enabled the British Empire to become the dominant global power during the nineteenth century. Since then the global spread of industry and the decline of this Empire has led to the increasing power and wider availability of weapons for many armed forces.

The international competition in the development for more destructive weapons increases the threat that powerful nations pose to each other and weaker countries. The danger of the ambitions of potential aggressors continues to influence the resources directed to maintaining a status quo that contains most conventional warfare. The quest for greater security for all is aided by efforts from the United Nations. Despite many limitations placed on this institution it does provide a centre for numerous organisations aspiring to direct global efforts aimed at solving the problems continually challenging many areas of humanitarian, ecological and economic activity.

The U.N. meditation room symbolises the occult 'New Spirituality' guiding this institution. The philosopher Teihard de Chardin's beliefs about humanities collective consciousness 'viewed the U.N. as the progressive institutional embodiment of his philosophy.' A famous quotation often attributed to him advises us that, 'We are not human beings having a spiritual experience. We are spiritual beings having a human experience.'

Epilogue

Cycles of Correspondence

The imagination of the early toolmakers who crafted flint into an increasing variety of cutting tools initiated a process the modern world depends on. Each new generation of jet airliners depends on the imagination of engineers to refine their designs and meet the modern standards safety demands. Time has not destroyed the correspondence between these two activities.

The way human potential is released to enhance the experiences of life is analogous to the manner plants express their potential by growing and releasing their own seeds, and possibly spread their species across the countryside. The annual cycle of the seasons illustrates an archetypal process that shepherds each generation of seeds to feed the world. This depends on Creation's use of the life force during spring, summer, autumn and winter.

It is a strange idea to think that reversing the process so that winter, autumn, summer and spring follow each other can also be a creative process. This thought is like reading a familiar story backwards and finding it too makes sense. This book describes how precession tells such a story. We will finish this book by looking at this remarkable idea in more detail.

From Table 10 we consider the twelve themes and their keywords for change related to the twelve constellations of the zodiac. We recognise that the physical meanings originated from our environment and were projected onto the heavens to create a calendar for our ancestors to use and better understand and manage time.

The order in which we will consider these cycles will not have the traditional start since this astronomical basis does not fit well this work's humanistic perspective. We will start with the month of Scorpio, the time when the life force is released.

Table 10's themes and keywords describe the qualities of physical, emotional, mental and spiritual dimensions of experience expressed by human consciousness. The evidence found about past spiritual achievements allows us to compare the practical potential of the precessional cycle against the normal zodiac cycle. When we consider these cycles we will note in each case the results human **strength** may achieve.

We will first look at the twelve spiritual qualities in the order of the cycle of the seasons. When we begin with Scorpio we know Sagittarius is next then Capricorn, Aquarius, Pisces, Aries, Taurus, Gemini, Cancer, Leo, Virgo and Libra.

With this sequence, we will describe the whole cycle in a strange sentence. It begins when power is *reborn*; it expands existing *beliefs* to control the *sacrosanct ideals* that evolve non-separation to *unity* with the **strength** of *will* to grow from fertile **aspirations** that *inspire* to meet with *blessings* of love from a radiant *self-awareness* with the purpose to *purify* relationships of *goodwill* so that power can be *reborn*.

In the second case, the order in which we use the same set of words is reversed. We know that in each humanistic age the spiritual keywords from Table 10 describe the primary motivating quality. When these words are used in this way, another strange sentence arises.

This sentence describes the terrestrial potential of the spiritual precessional cycle. This process begins with Scorpio, which is then followed by Libra, Virgo, Leo, Cancer, Gemini, Taurus, Aries, Pisces, Aquarius, Capricorn and Sagittarius.

Following this order, the themes and keywords describe the *rebirth* of power to relate *goodwill* of purpose to *purify* a radiant *self-awareness* that *blesses* with love to *inspire* people to make connections with fertile *aspirations* that promote a **strong** *will* where the non-separation of **unity** evolves with *ideals* to control *sacrosanct beliefs* about the power of *rebirth*.

In the first sentence, beliefs arise from the experience of goodwill. In the second sentence, beliefs develop in harmony with sacrosanct ideals.

Perfection's Perception

In nature we know the cycles she spins,
The web of perfection the prize that she wins.
The continuing saga of Sun, wind and rain,
Aeons of time in endless train.

The world now revealed to man's finite mind
Shows patterns of purpose in all that he finds.
Since mesmerised by our five senses,
We are attracted away from where the true scent is.

The scales often used for measuring good,
Cover our eyes try as we would.
The laws that are followed can be so very subtle.
Not knowing our own we ensure our rebuttal.

Not to remain as one of the haunted
The mind may focus on spirit undaunted.
The process needed to break loose from the past
Requires its forgiveness for improvements to last.

Perfection's shy veil shields all within sight,
For whatever went in the outcome is right.
The problem this poses, try as one might,
Can only be solved with clear inner light.

That justice is there clothed in pain and disease,
Few are the minds to perceive this with ease.
From within each illusion strange patterns emerge
The deeper truths may be released with a surge.

When the dance of imbalances is by beauty graced
The threads of all life become interlaced.
The extremes of pain are by love replaced
To reveal what past fears had formerly lain waste.

The new perspectives needed are not easily found,
with head in the air and feet on the ground.
If wisdom graces us to secure such huge gains
The insights given come not from Earth's plane.

Andrew Bradbery

Appendix

In Chapter 12, the historical record was aligned with the cycle of precessional symbolism. The correlation between many religious events leading to and including the reign of Rameses II with the precessional summer solstice is selected to illustrate the alignment with 1260 BC. These events are chosen because the summer solstice is the widely appreciated peak in the annual cycle of the seasons and the famous reign of Rameses II in Egyptian history is also widely recognised. Many other people famous for their contribution to cultural progress during recent millennia were used to check the alignment of the historical record with the precessional cycle. Their accomplishments correlate with the symbolism of the degree aligned with each life.

To illustrate the technique used for these correlations we will look at the symbolism for the degrees before and after the summer solstice. This corresponds with the 72 year degree periods before and after 1260 BC. We will use the idea that every degree's archetypal essence is immanent for 72 years as described in Dane Rudhyar's book *An Astrological Mandala*. The symbolism of the degrees in quadrature to this time, further validate the events in the context of the precessional cycle.

Precession requires that we consider the 1st degree of Cancer before the 30th degree of Gemini. In this order, their symbolism when interpreted correlates with cultural transitions arising from a *change in loyalty* followed by the welcome acceptance of change after the *display of natural and or cultural excellence.*[1]

During this period of Egyptian history, several changes in cultural loyalty were earlier seen to have spiritual agendas.

1. The worship of sun disc (the Aten) at Karnak, the traditional home of the national god Amun, created a revolution.

2. The divinity of Amenhotep III was worshipped by his son

[1] Dane Rudhyar, *An Astrological Mandala,* p.108-109

Akhenaten. This was heretical if it took place *before his death*.

3. The worship of the Aten was established as the new national religion in place of the Amun god with its many temples and priesthood traditions.

4. The new city of Akhetaten at el-Amarna replaced Thebes as the capital city of Egypt. All its citizens were able to worship Akhenaten's monotheistic god in public ceremonies.

5. The Amun temples were closed and their riches taken for the Aten cult. The Amun priesthood was sacked.

6. Akhenaten's son Tutankhaten (living image of the Aten) changed his name to Tutankhamun (living image of Amun). This symbolised the pharaoh's change of allegiance and the final point of return for Egypt to follow their religion of the god Amun-Ra.

7. The superior display of spiritual power by Moses against the Egyptian priests led to Ramesses II losing his Hebrew slaves.

8. After this failure, the Amun priesthood retained their religious power by adopting Amun-Ra as their supreme god.

This succession of events describes different times when the securing of lasting spiritual power waxed and waned.

The Exodus of the Israelites out of Egypt gave them the freedom to follow the monotheistic religion Moses introduced. His leadership *displayed the cultural excellence* that depended on the use of spiritual intelligence, and contributed to the spread of monotheism.

We can further appreciate these cultural dynamics when we consider these two archetypes in their annual order; *displays of local excellence* lead to a *change in loyalties* and enhance the traditions that supported them. Reversing this order, any doubts to local change are reduced when foreign cultural *displays of wanted excellence* are exhibited. Then, entrenched traditional *loyalties change*. This reversal fosters a global culture depending on the best practice suited to local conditions. Instead of cultures becoming more entrenched, they become more diverse.

Selected Bibliography

Aldred Cyril. *The Egyptians.* London, Thames and Hudson, 1992.

Ahmed Akbar. *Discovering Islam.* Routledge, 2002.

Anthony David W. *The Horse, The Wheel and Language.* Princeton University Press, 2007.

Bloom and Blair. *Islam – Empire of Faith.* BBC Worldwide Ltd., 2001.

Brady, Bernadette. *Brady's Book of Fixed Stars.* New York, Weiser, 1998.

Brown, Peter. *The Rise of Western Christendom.* Oxford, Blackwell, 1997.

Burrell, Roy. *The Greeks.* Oxford University Press, 1991.

Bauval & Hancock. *Keeper of Genesis.* London, Heinemann 1996.

Baring and Cashford. *The Myth of the Goddess.* London, Arkana, 1993.

Bottero Jean. *Religion in Ancient Mesopotamia.* London, The University of Chicago Press, 2001.

Carlyle, Thomas. *On Heroes and Hero Worship.* London, Oxford University Press 1974.

Chidester, David. *Christianity a Global History.* London, Penguin Press, 2000.

Childre and Martin. *The HeartMath Solution.* New York, Harper Collins, 1999.

Clarke, Dr. Peter B. *The Worlds Religions.* London, Reader's Digest Association, 1993.

Clayton, Peter A. *Chronicle of the Pharaohs.* London, Thames and Hudson, 1994.

David and Goudie. *The Oxford Companion to Global Change,* Oxford University Press 2009.

Cunliffe, Barry edited. *The Oxford Illustrated History of Prehistoric Europe.* Oxford, Oxford University Press, 1997.

Davis, Norman. *Europe A History.* Oxford, Oxford University Press, 1996.

Duncan, David. *The Calendar.* Forth Estate Ltd, 1998.

Frazer J.G. *The Golden Bough.* London, Octopus Publishing, 2000

Gimbutas, Marija. *The Language of the Goddess.* New York, Thames and Hudson, 2001.

Gleick, James. *Chaos : making a New Science.* New York, Viking Penguin, 1987.

Gonzalez, Justo, *The Story of Christianity, Vol 1.* Harper Collins, New York, 2010.

Grant, Michael. *Myths of the Greeks and Romans.* New York: Penguin Meridian, 1995.

Grant, Michael. *Greeks and Romans, A Social History.* London, Weidenfeld and Nicolson, 1992.

Grof, Christina and Stanislav, *The Stormy Search for the Self,* New York, Tarcher/Putnam/Penguin. 1990.

Grof, Stanislav. *The Cosmic Game.* Newleaf, Dublin, 1999.

Grimal, Nicolas. *A History of Ancient Egypt.* Blackwell, Oxford, 1992.

Graves-Brown, Carolyn, *Dancing for Hathor – Women in Ancient Egypt,*

Hancock, Graham and Robert Bauval. *Keeper of Genesis.* Heinemann, London, 1966.

Harris, D.R.& G.C. Hillman edited, *Foraging and Farming.* London, Unwin Hyman 1989

Haywood John. *The Ancient World.* London, Quercus Publishing. 2010.

Haywood John. *The Atlas of Past Times.* Worcester, UK, Sandcastle Books Ltd. 2006.

Hope, Murry. *The Sirius Connection.* Shaftesbury: Element, 1996.

Hutton, Ronald. *The Pagan Religions of the Ancient British Isles.* Oxford, Blackwell, 1999.

Ian Shaw edited. *The Oxford History of Ancient Egypt.* Oxford University Press, 2003.

James Gleick. *Chaos - Making a New Science.* New York: Viking Penguin, 1987.

Johnstone, W. *Exodus - Old testament Guides.* Sheffield Academic Press, 1999.

Johnson, Paul. *The Civilization of Ancient Egypt.* Seven Dials, London, 2000.

Kerenyi, C. *The Gods of the Greeks,* Thames and Hudson, 1988.

Kriwaczek Paul. *Babylon Mesopotamia and the Birth of Civilization.* London: Atlantic Books, 2010.

Kritzinger, Ann. Bring it to Book: London, Scriptmate Editions, 1997.

Laney, J Carl. *The Concise Bible Atlas.* London, Marshall Morgan and Scott, 1977.

Leick Gwendolyn. *Mesopotamia the Invention of the City.* London, The Penguin Press, 2001.

Lemesurier, Peter. *The Great Pyramid Decoded.* Shaftesbury, Element, 1996.

Lewis-Williams, David. *The Mind in the Cave.* London, Thames and Hudson, 2002.

Loewe, Michael. *The Pride that was China.* London, Sidgwick and Jackson, 1990.

Mithen, Steven. *After the Ice – A Global Human History.* London: Weidenfeld & Nicolson, 2003.

Mithen, Steven. *The Prehistory of the Mind.* London, Thames and Hudson, 1996.

Mitton, Jacqueline. *Dictionary of Astronomy.* London, Penguin, 1998.

Morris, Ian. *Why the West Rules – For Now*: London, Profile Books, 2010.

Mellars, Paul. *The Neanderthal Legacy.* New Jersey, Princeton University Press, 1996.

Moynahan, Brian. *The Faith – A History of Christianity.* Aurum Press: London, 2002.

Oken, Alan. *As Above So Below.* New York, Bantam, 1973.

Quirke & Spencer. *The British Museum Book of Ancient Egypt.* British Museum Press, 1993.

Quirke, Stephen. *Ancient Egyptian Religion.* British Museum Press, 1992.

Paul Gepts, Thomas Famula, Robert Bettinger, Stephen Brush, Ardeshir Damania, Patrick McGuire and Calvin Qualset edited, *Biodiversity in Agriculture.* Cambridge University Press, 2012.

Paungger & Poppe. *The Art of Timing.* Saffron Walden, 2000.

Pollack, Rachel. *The Body of the Goddess.* Shaftesbury, Element Books, 1997.

Powell & Threadgold. *The Sidereal Zodiac.* The American Federation of Astrologers, 1985.

Powell Robert. *History of the Zodiac.* Sophia Academic Press, 2007.

Reeves, Nicholas. *Egypt's False Prophet – Akhenaten.* London, Thames and Hudson, 2005.

Robbins Ph.D., F. E. edited and translated, *Ptolemy Tetrabiblos.* Massachusetts: Harvard University Press, London, William Heinemann, 1980.

Roberts, J.M. *The Penguin History of Europe.* London: Penguin, 1996.

Rudhyar, Dane. *The Pulse of Life.* Shamballa, Berkeley & London 1970.

Rudhyar, Dane. *An Astrological Mandala*, Vintage Books, Random House, New York, 1974.

Saggs, H. W. F. *Peoples of the Past – Babylonians.* British Museum Press, 2000.

Smart, Ninian. *The World's Religions.* Cambridge University Press, 1992.

Shaw, Ian. *The Oxford History of Ancient Egypt.* Oxford University Press, 2000.

Tarnus, Richard. *The Passion of the Western Mind.* New York: Ballantine, 1993.

Tarnus, Richard. *Prometheus the Awakener.* Oxford: Auriel Press, 1993.

Ulansey, David. *The Origins of the Mithraic Mysteries.* Oxford University Press, 1991.

Wilber, Ken. *No Boundary.* Boston: Shambhala, 2001.

Wilkinson, Toby. G*enesis of the Pharaohs.* London: Thames and Hudson, 2003.

Wilkinson, Toby. *Lives of the Ancient Egyptians.* London: Thames and Hudson, 2007.

Wilson, Colin. *The Occult.* St Albans: Mayflower, 1972.

Winstone, H. V. F. *Uncovering the Ancient World.* London: Constable, 1985.

Xinzhong Yao. *An Introduction to Confucianism*, Press Syndicate, the University of Cambridge, 2000.

Tables and Fig Numbers listed by Chapter

NOTE: entries in *italics* indicate keywords for consciousness attributes.

CPSIA information can be obtained
at www.ICGtesting.com
Printed in the USA
LVOW13*0519270617

539511LV00006B/20/P